L. E. Weiss

The Minor Structures of Deformed Rocks

A Photographic Atlas

With 203 Plates

Springer-Verlag
Berlin Heidelberg New York 1972

Professor Lionel E. Weiss

Department of Geology and Geophysics
University of California
Berkely, California/U.S.A.

ISBN 3-540-05828-1 Springer-Verlag Berlin Heidelberg New York
ISBN 0-387-05828-1 Springer-Verlag New York Heidelberg Berlin

Printed in Germany. Typesetting, printing and bookbinding: Universitätsdruckerei H. Stürtz AG, Würzburg

Preface

Perhaps the most striking impression gained by a geologist during twenty years of field work with deformed rocks in many parts of the world is that the minor structures in these rocks are surprisingly uniform in their properties and restricted in their variety. In fact, a relatively short and simple list can include all structures which are both of common occurrence and of use to the structural geologist in his attempts to understand the structure and evolution of a deformed geologic body.

The photographs in this book have been selected to illustrate as clearly as possible the obvious characteristics of the common minor structures of deformed rocks. I have tried to make this selection truly representative in spite of two factors that have affected the choice. First, I have included only photographs made during my own field work. No doubt the selection could have been greatly improved in both quality and variety by including photographs made by other geologists. Second, the photographs necessarily reflect my own interests and not those of every geologist concerned with deformed rocks. Consequently, for example, great emphasis has been placed on folds, whereas faults and joints are entirely omitted.

I have avoided a comprehensive description of the illustrated structures and presented instead a brief essay-like commentary. In the same vein, I have refrained from including a critical glossary, and have tried to use only those terms currently employed by most geologists. No particular classification of structures has been included, except that of the most obvious kind.

The main aim throughout has been merely to *illustrate* the common minor structures resulting from deformation. Other works exist in which attempts have been made to interpret these structures in the light of field, experimental and theoretical studies. However, it is difficult to write even superficially about such structures as foliations and folds, for example, without revealing one's own prejudices about their origin. In the introduction to the plates, therefore, some statements about the genesis of certain structures have been included. I hope that these ideas will not prove too unacceptable to most field geologists and will not obscure the main purpose of the photographs.

Many people have assisted me in the preparation of this book. In particular I thank the following: VOLKMAR TROMMSDORFF, University of Basel; KNUT HEIER, University of Oslo; H. B. WHITTINGTON, University of Cambridge; JOHN RAMSAY, Imperial College, London. Others, such as J. DONS, JOHN DEWEY, TIM HOPWOOD and PAUL WILLIAMS have provided me with material to photograph, or led me to it in the field. I am deeply grateful also to the John Simon Guggenheim Memorial Foundation of New York who made possible much of the necessary field work; and finally, I would like to thank Dr. KONRAD F. SPRINGER who has taken a personal interest in this book and assisted me greatly in its preparation.

Contents

Introduction

The earth scientist is concerned with physical and chemical phenomena occurring in bodies which vary greatly in size. The interests of a single geologist, for example, can range from the chemical interactions between microscopic mineral grains in a rock to the differential motions of lithospheric plates or other large tectonic units of the crust, hundreds or thousands of kilometers in extent. He can be concerned, in fact, with attempting to establish a relationship between phenomena of such disparate sizes.

In most branches of the science of solids, interests of the scientist are much narrower. The solid-state physicist, for example, generally concerns himself with interactions among atoms, fields and other phenomena in single crystals and with the physical properties these interactions express on a larger scale. He rarely concerns himself even with the properties of polycrystalline aggregates, leaving the study of the structure and properties of these to the materials scientist, the metallurgist, the engineer, and so on.

In structural geology, which in its widest sense includes also tectonics—the study of the larger building blocks of the earth—concern is necessarily with the structural properties of rock-building materials on a variety of scales. The simplest solid materials of the earth are mineral crystals, generally of very small size, arranged in the polycrystalline aggregates we know as rocks. Rocks vary widely in mineral content and internal structure and are themselves arranged in larger structural units. Sedimentary rocks, for example, occur in beds or strata; igneous rocks can occur in dikes, sills, plutons and so on. The field geologist tends to designate rock units as *formations* if they are separated from surrounding rocks by well defined and easily mappable surfaces. But for many purposes, still larger tectonic units must be designated in a somewhat arbitrary fashion, according to the aims of a particular geologic study. Thus fold mountain belts, stable platforms, continental shields and so on have been recognized in the continental crust which itself is fragmented to form parts of the largest presently recognizable tectonic units of the upper earth—the lithospheric plates. Structural units of these kinds recognized by the geologist range in size from about 10^{-4} mm, for the diameter of crystals of clay minerals in a finegrained shale, to about 10^{9} mm for the diameter of the largest lithospheric plates. At 2×10^{3} mm, almost midway in size between crystals in the finest-grained rocks on the one hand, and the whole earth on the other, stands the average geologist.

Geology began and continues to this day as a dominantly field science. In recent years, indirect geophysical methods using the magnetic, gravitational, thermal and other physical properties of the earth, as determined by sensitive instruments, have contributed otherwise unobtainable information to our knowledge of the earth as a whole. Likewise, marine geology with its indirect explorations of the ocean floors has helped form recent ideas of global tectonic processes and increased our understanding of the manner of interaction of crust and mantle, as pictured in the theory of plate tectonics. But the main source of raw data for most geologists working in the continental crust remains direct observation of rocks in the field. The geologic phenomena similar in size to the geologist himself, namely, those that can be directly examined in the field, in hand specimen and small exposure, remain—and are likely to remain for a long time—central to geologic investigation. Without doubt, a modern geologic study of a region requires that the structure and composition of rocks be examined in the laboratory with the advanced techniques now available; and also, that larger-scale syntheses be made by the combination and interpretation of all possible information. But even in these studies field observations remain central: the choice of hand specimens and samples for laboratory analysis is made in the field, and the accuracy of any large-scale synthesis generally depends upon the careful gathering and interpretation of field data.

This central role of field study is very evident to the structural geologist. Many of the present-day notions about large-scale tectonics rest upon direct examination of geologic "structures" in the field. What is a geologic structure? There is a dichotomy in the use of the term structure even in everyday language. In conventional usage the term can denote some configurational entity, such as, for example, a bridge or a building. In this context one speaks generally of "a structure." The term is used also to denote the internal configuration or organization of an extended entity of some kind. In this context one speaks of "the structure" of, for example, a crystal, music, or the federal government. This dichotomy persists into geologic usage where one may speak of *a* structure, such as a syncline, or of *the* structure of a particular region, implying its internal configuration.

In the title of this book emphasis seems to be placed upon the first of these meanings. There is, however, no clear dividing line between them and none is intended here. Indeed, what is a discrete structure in a rock body on one scale can be viewed as one element of its internal structure on another. In the field a geologist observes directly those structural features of dimensions similar in magnitude to his own. Such features fall into both of the above categories. A single small fold, for example, is a discrete structure, whereas the foliation of a schist is a pervasive structure defined generally by the preferred orientation of platy minerals of microscopic size. But on a larger scale, the small fold may equally properly be viewed as a pervasive structural feature in a large body containing thousands of similar structures in a state of preferred orientation.

Structures such as small folds and foliations, which can be examined directly and have directional geometric properties which can be established simply in the field, are very important to the structural geologist. They provide data indispensible to any regional structural synthesis and careful study of these and other directly observable structures of sedimentary, igneous or metamorphic origin yields information on a variety of geologic processes. Thus structures of "middle size" in the earth (or *mesoscopic* structures, as they have been called), assume a unique importance in geologic studies of all kinds. Some of them such as folds, occur on both larger and smaller scales with much the same properties; others, such as cleavages and lineations, are in effect confined to this scale—on smaller scales we must view them in terms of the microscopic features of fabric and texture they express, and on a larger scale we can examine only their orientations over large regions.

In accordance with current practice in many English-speaking countries, structures which can be examined directly in the field in hand specimens or exposures are here termed *minor structures;* and with few exceptions the photographs which follow are of such structures observed in the course of routine field work.

It is conventional and useful to designate the structures of rocks either as primary—those resulting from sedimentary or igneous processes, or as secondary—those formed in pre-existing rocks as a result of metamorphic and similar processes. In a strict sense, deformation is a part of metamorphism: but not all deformed rocks are truly metamorphic, and not all metamorphic rocks are deformed. Thus, for example, many otherwise unmetamorphosed sedimentary rocks are strongly folded, and some contact metamorphic rocks contain the most delicate primary structures preserved in an undistorted state.

In most situations the experienced field geologist is able to distinguish without question structures formed by deformation from those resulting from other processes. There are, however, instances where this distinction is equivocal: some primary structures resulting from slumping or flow in unconsolidated sediments, lava, or magma simulate structures formed by flow in the solid state—particularly where the primary structures have been markedly modified by later deformation of a tectonic nature. Evidence suggests also that some structures generally viewed as strictly secondary, such as, for example, slaty cleavage, may develop, at least incipiently, in unconsolidated sediments. But most geologists agree both on which structures should be identified as " deformation structures " and on a list of the commonest types.

Given, then, that most structures resulting from deformation of solid rock under metamorphic and nonmetamorphic conditions can be identified in the field, how does a geologist study them, and what information of wider geologic importance can be obtained from such study? Several publications of a general kind have addressed themselves to this problem (see general references at the end of the volume) and hundreds of papers* have considered the properties and genesis of these structures from a variety of view points. These and other works have shown how the nature and orientation of directional minor structures resulting from deformation can be used both as a guide to the regional structure of large bodies of deformed rock and also to the evolution of such bodies—even where several episodes of deformation are involved.

There is nowadays a tendency for structural geologists not to be satisfied with a straightforward determination of structure in a deformed geologic body: other questions are asked, such as the following:

1. How and through what stages did the deformed body evolve from an initial undeformed state?

2. What were the patterns and magnitudes of local and regional displacement and strain involved in the episode or episodes of deformation? How can these be related to larger motions in the crust and upper mantle?

3. By what physical mechanism did the rocks flow or fracture? What was the rheologic state of the rocks? What are the physical and chemical processes by which the common structures such as foliations, lineations and folds are formed?

4. What were the physical conditions at the time of deformation, and how did they vary with time?

Few of these questions can be answered for any particular body of deformed rock. But detailed field investigation continues to provide data bearing on some of these problems, and also forms the basis for the design of worthwhile experimental and theoretical investigations essential to the solution of others.

Perhaps the greatest problem facing the structural geologist is to be sure that the questions that he asks are realistic. For example, with ideas colored by a knowledge of the theory of elasticity, some geologists attempt direct interpretation of deformation structures in terms

* Rather than attempt to prepare a comprehensive bibliography of realistic length from the large and rapidly growing literature on the structure of deformed rocks, I include a short list of what I consider to be key books and papers at the end of the volume. Lengthy bibliographies can be found in some of these works, and new papers of importance appear constantly in the current journals.

of stress. Such interpretations are generally possible for elastic deformations where strains of very small magnitude are to a first approximation functionally related to stresses. But where deformation is permanent and of large magnitude—as is true in deformed rocks—the state of strain is generally not in simple relationship to any particular state of stress, regardless of the rheologic state of the rocks. In all types of mechanical behavior to be expected in rocks a particular state of strain is reached along a path from some initial state. At each point along this path there can be some relationship between the instantaneous stress state and some aspect of the incremental strain and there can be some material coefficient, such as, for example, a coefficient of viscosity, which determines this relationship. Several considerations suggest, however, that the relationship varies with the strained state and in a nonlinear way. First, mechanical anisotropy of a rock as expressed by its internal microstructure or fabric appears to change progressively with state of strain: rocks with planar layers become folded, foliations and lineations develop. Such changes must be accompanied by corresponding changes in material properties. Second, the rheologic state of a rock, which governs the constitutive equation for flow, is generally sensitive to the changes in physical conditions of temperature, confining pressure, presence of pore fluids, and so on, to be expected along any deformation path. Third, the distribution of surface and body forces must change along with changes in the tectonic environment. In this fashion changes in the state of stress can occur independently of changes in material properties.

The above factors suggest that it may not be realistic to interpret most deformation structures—including microfabrics—in terms of states of stress. A few exceptions to this general conclusion occur in situations where the aggregate strain represented by deformation structures is very small—close to that for an elastic deformation. Thus some fracture systems in which small displacements have occurred on widely spaced surfaces are susceptible to stress analysis, as are certain thin zones of plastic yielding such as mechanical twin lamellae and deformation bands in crystals and, possibly, some kink bands in rocks—provided that a directional yield criterion exists as a basis for analysis.

On the other hand, determination of the state of finite strain from deformed structures of known initial configuration is independent of any prior knowledge of material properties or of the path by which the strain developed, and such studies are becoming increasingly common and successful among structural geologists.

At the time of writing, the mode of origin of almost none of the common minor structures of deformed rocks is fully understood. Recent theoretical and experimental studies have begun to attack the problems of folding and development of foliations, and much progress has been made also in understanding the roles of phenomena such as crystal plasticity and creep in the natural flow of rocks. But few general principles have emerged.

Most field geologists, however, are not and should not be primarily concerned with the mechanical problems of rock flow. The main tasks in field study is to prepare a map and a three-dimensional structural interpretation of a rock body. These tasks can be completed if necessary with little understanding of the physical and chemical processes involved in the flow and fracture of rocks. However, if the evolution of a deformed body from an earlier state or states is to be understood, some attention should routinely be paid to the strains, displacements and rotations which have occurred, both on the local and on the regional scales. Because this generally can be done without appealing to mechanical principles, more and more geologists working primarily in the field are showing interest in the kinematic significance of common deformation structures and the interpretation of these, at least qualitatively, in terms of strains and displacements. It is partly which such geologists in mind that the following brief commentary on the plates has been included. Wherever possible, simple interpretations of structures in kinematic terms have been indicated. The discussions are far from exhaustive; but I hope that the comments will give to geologists unfamiliar with and confused by deformation structures some feeling for the information of general geologic interest they can provide. Other geologists, perhaps more familiar than I with the structures illustrated, will, I hope, gain some pleasure from merely looking at the photographs.

Introduction to the Plates

There appear to be few more hazardous occupations for the structural geologist than to attempt classification of the common minor structures of deformed rocks.* The main difficulty in arriving at a useful classification seems to be the lack of a clear-cut basis on which to classify. The relationships between even the common types of structure are poorly understood, and a purely descriptive basis of the kind favored by most geologists generally results in the separation into different classes of structures of very similar origin because of differences in appearance from place to place. Likewise, a purely genetic basis fails because we do not know how most structures are formed.

For these and other reasons, including a desire to be as brief as possible, no attempt is made here to include a formal classification of minor structures or a critical glossary of classifications and terms in general use. Instead, structures are loosely grouped into gradational classes of a most conventional kind, largely on the basis of the kind of information they provide the field geologist. This grouping is reflected in the arrangement of the plates, and it reflects also some of my own prejudices about the genesis and geologic significance of these common structures.

No attempt is made to illustrate all possible types of minor structure—some types have been omitted altogether. Only the common structures are included and it is assumed that the reader is already familiar with these structures in a general way.

In the captions of the plates the terms "horizontal" and "vertical" are used respectively for across and up and down the page. Similarly, "dipping left" means crossing the page from upper right to lower left. The actual field orientations of the structures are not always apparent. Scale is generally indicated by the inclusion in the photograph of some familiar object (generally a hammer or a coin). Otherwise, dimensions are given in the captions.

Specific plates are frequently cited in the following discussion as examples of particular structures. Other examples of the same structure are commonly to be found in a number of other plates, and many plates illustrate structures of several types.

* Not that many of us have been turned aside from this task by its dangers; one recent textbook, for example, lists almost two hundred different terms that have been used in the classification of folds.

Planar Structures (Plates 1 to 51)

Some of the earliest workers with deformed rocks recognized that rock cleavage and similar pervasive planar structures resulted from deformation. Many terms have been used over the years to describe different varieties of these structures and there is yet little general agreement in the use of these terms. My personal prejudice, adopted here, is to call all such structures *foliation* (a term pleasingly parallel to *lineation* for pervasive linear structures) regardless of detailed characteristics. Under this general head a number of commonly recognized varieties can be listed as: slaty cleavage (Plate 12); crenulation (or strain-slip) cleavage (Plate 25) fracture cleavage (Plate 2); schistosity (Plate 34); transposition foliation (Plate 27); axial plane foliation (Plate 23); and the coarser lamination and layering of some schists and gneisses (Plate 46)—to which the term foliation is restricted by some geologists.

All of these structures are in fact some combination of relatively few microscopic structural features, such as planar preferred orientation of inequant mineral grains and planar segregations of mineral grains into layers or lenticles. Also important are parallel sets of closely spaced fractures and attenuated limbs of microscopic folds or crenulations. The obvious differences between foliations seem to reflect several factors such as, for example:

1. The initial composition and structure of the rock;
2. The physical conditions under which the foliation formed;
3. The stage to which the foliation has developed in response to deformation and mineralogical and textural reconstruction during metamorphism.

In the field it is convenient to divide foliations into two broad groups. First, foliations developed together with continuous planar structures, such as bedding or an earlier foliation, and second, more strongly developed foliations not obviously associated with any continuous earlier planar structures. Structures of the first kind (Plates 1 to 28) are easily seen to be of secondary origin. Where folds in the earlier surfaces are present, the foliation is generally parallel or otherwise symmetric to the axial surfaces of these folds. Structures of the second kind (Plates 29 to 51) as less easily identified as secondary structures. Lamination, layering and segregation of mineral grains of the same type can occur parallel to the foliation surfaces in a manner that can simulate the bedding of sedimentary rocks, at least locally. The term *bedding foliation* has been applied to some structures of this kind, and indeed, some of them appear to be relatively undisturbed bedding (Plate 48B). Others, however, are clearly of secondary origin as can be demonstrated by local survivals of folded and disrupted fragments of earlier layers (Plates 41 and 42). These fragments, if plentiful, can form overlapping lenticular or elliptical bodies typical of some types of *mélange* (Plates 31 to 33).

Study of foliations of various kinds suggests that they are all to some degree "transposition foliations" in that they represent different stages in different rock types of the disruption and eventual replacement of one planar structure by another. The earlier stages of this process are preserved in foliations of the first kind mentioned above. The common structures such as slaty cleavage, crenulation cleavage, "axial-plane foliation," and so on, are indicative of incipient transposition of earlier surfaces—bedding or foliation—into a later foliation. At such an early stage in foliation-development continuity of the earlier surfaces is preserved and the relationship of the developing foliation to other structures such as folds, boudins and lineations is clear. At later stages of foliation-development, boudins and intrafolial folds (fold boudins) can survive as indications that transposition has occurred; but continuity of the original surfaces is lost. The final stages can yield an apparently uniformly foliated or layered rock in which all obvious traces of earlier surfaces are, at least locally, destroyed (Plates 29, 34, 36, 37, 46 and 47).

The importance of transposition of planar structures in the evolution of most deformed geologic bodies cannot be overemphasized. By this process the initial stratigraphic continuity of beds becomes disrupted and replaced by a mechanical planar structure which itself can become folded and partly or completely transposed in later episodes of deformation. The actual appearance of a transposition foliation in the field depends upon a variety of factors, the most important of which may well be the initial nature of the rock. In a rock initially containing isolated layers of characteristic composition enclosed in a different matrix—such as, for example, thin layers of quartzite in shale—transposition results in the folding and subsequent boudinage of the layers to give lenticular fragments which "float" in a foliated matrix. Such structure is observed commonly in *mélanges* (Plates 30 and 31). More massive rocks acquire a more regular type of foliation; and the disrupted fragments of older layers need not be obvious (Plates 16, 17, 29, 45 and 47). Transposition of a well developed earlier foliation, such as a schistosity, generally takes place by the progressive development of a strain-slip or crenulation foliation. Thin zones of strain defined by the attenuated limbs of crenulations in the earlier foliation become wider and more closely spaced as deformation proceeds (Plates 21, 22, 25, 27 and 28). These zones are generally sites of mineral growth and are not simply mechanical slip surfaces.

The effect on foliation-development of initial variation in rock composition is most obvious in incipiently foliated or cleaved well bedded sedimentary rocks. The character and orientation of the cleavage can be different in layers of different composition (Plates 3, 4,

14 and 19). Such change in orientation across boundaries between beds of different composition is well known as "cleavage refraction."

There is some tendency for rocks to become structurally and compositionally "homogenized" by foliation development, at least where no pronounced metamorphic differentiation occurs. Successive development of foliations in several episodes of deformation can turn an initially bedded sequence into much more compositionally uniform schist or gneiss.

The development of foliation, then, is intimately bound up with the development of folds and boudins (see below). Foliations are commonly subparallel to the axial planes of associated folds (Plates 7A, 11B, 13B, 21 and 26A) or symmetrically arranged about them in divergent or convergent fans (Plates 13A, 20A, 85 and 86A). Boudins tend to result from extension in the plane of associated foliation and flattening normal to it (Plates 145 to 148). These phenomena suggest that most foliations are formed in response to a regional "flattening" or shortening of a body normal to foliation. The old controversy about foliation involving shear (that is, simple shear on the foliation) versus flattening (pure shear with the foliation as a principal plane) does not seem relevant nowadays. Distortions determined from structures such as folds, boudins and deformed primary structures associated with foliation suggest that elements of both "theories" are valid locally. Local slip on foliation surfaces is hard to deny and it seems equally clear that no foliation can form without major shortening normal to the plane of the foliation.

Linear Structures (Plates 52 to 66)

Although as common a deformation structure in rocks as foliation, lineation has received far less attention from geologists. Some of the earliest geologists working in the Swiss Alps and the Scottish Highlands observed the very obvious regional lineations in these mountains and even commented on their probable significance. But for a long period lineations were all but ignored; and when attention turned to them again, forty or fifty years ago, they became a center of controversy not yet resolved in the minds of some geologists.

In many ways it is difficult to treat lineations separately from foliations. Both of these structures are expressions of a systematic microstructure in rocks, and in some rocks exactly the same microscopic feature—such as, for example, the preferred orientation of inequant mineral grains—defines both structures. There is a complete gradation between foliated rocks with no lineation and lineated rocks with no foliation. Most deformed rocks are both foliated and lineated to some degree.

The microscopic features of deformed rocks expressed on a larger scale as lineations visible in the field are few in number. The commonest are, first, preferred orientation of inequant grains, aggregates of grains or fragments; second, the hinges of microscopic crenulations in associated foliations or other planar structures; and third, the intersection of more than one set of pervasive planar structures—a lineation is present on each set of surfaces in the direction of its intersection with any other. These features can occur in a variety of combinations to form linear structures of widely different appearance.

Fortunately, relatively few terms describing lineations have achieved widespread usage in English. Most geologists seem to distinguish two ideal types; first, what might be called a *mineral lineation* or "streaking" defined by the preferred orientation of inequant grains and other bodies (Plates 53 A, 57 B, 58, 59, 60, 61 and 62 A); and second, *crenulation lineation* defined by the hinges of microscopic folds or crenulations (Plates 52, 53 B, 54, 56 and 57 A). Lineations of a common regional type (Plate 55) are defined by both features; and in the field give a ribbed or grooved appearance in some layers and a clearly crenulated appearance in others. A third type of *intersection lineation* is also common, especially in folded rocks with axial-plane foliations; and is gradationally related to the other two.

In simple cylindrically folded rocks, the common type of lineation is subparallel to the fold axis. Indeed, the hinges of very small folds themselves define a coarse type of lineation (Plates 56 and 63). In the hinge regions of very large folds, a coarse type of lineation termed *mullion structure* can be locally developed (Plates 63 to 66). There seems to be no real difference between this structure and the finer lineations, except the scale of development. Lineation of other types, as well as mullion structure, seem to be better developed in hinges of folds than on limbs, and regions of intensely linear structure in bodies of deformed rocks commonly can be identified as the hinges of large folds, even where partial transposition of the folded surfaces has occurred. Throughout many deformed bodies the orientation of lineation and mullion structures as measured in the field is a clear guide to the orientation of regional fold axes (Plates 58 and 60).

The relationship between common types of lineation and regional strain is by no means always clear. No lineations have been made experimentally under realistic physical conditions and most ideas about the kinematic significance of lineation came from indirect evidence. Crenulation lineations seem to have much the same significance as folds. They occur commonly in association with incipient or well developed crenulation or strain-slip types of foliation, the lineations being defined by the crenulations of the earlier foliation and by the line of intersection of the earlier and later foliations. In fact, all foliations of the crenulation type must be associated with a coeval crenulation lineation. Mineral lineations can occur parallel to associated crenulation lineations; but there is good evidence from deformed primary structures, such as fossils and pebbles, that some mineral lineations form in the direction of greatest extension in a rock body. In Plates 183 and 184, for example, faint lineations can be clearly seen in the direction of greatest extension of the fossils, and in Plate 155 a prominent mineral lineation is shown coinciding with the extension direction determined from associated boudins.

The extension direction in many deformed rock bodies appears to be along the regional fold axis; but this is not always so. In some folded rocks with axial plane foliation the direction of greatest extension lies in the foliation but obliquely inclined to the fold axis.

Lineation, then, does not have a unique significance in deformed rocks. It has been interpreted in a variety of ways, including the slip direction where simple shear occurs on a foliation surface. Most lineations seem to be either in the direction of maximum finite elongation, or in the direction of fold axes or foliation intersections. In some rock bodies these directions coincide.

Folds (Plates 67 to 140)

Folds are the most obvious and abundant evidence of permanent deformation in bedded, layered or foliated rocks. They occur in both metamorphic and nonmetamorphic rocks and range in size from microscopic crenulations—generally classed as lineations—to nappe structures many kilometers in amplitude. Because of their abundance and importance folds have received attention from geologists of all kinds over a very long period, and a complex and confusing terminology defining hundreds of varieties of folds has grown up. Few of these terms seem really important in the light of recent detailed geometric studies of folds—especially minor folds, our present subject.

An important notion to emerge from geometric study of folds is that of a cylindrical fold. Many segments of minor folds observed in the field fall into this category, and the shape of such folds is accurately shown only in true profile—in the plane normal to the fold axis. In nature, joints are commonly subparallel to this plane and the true profiles of many minor folds can be studied and photographed on these surfaces. Most of the folds shown in the plates are in true profile or almost true profile.

The classic geometric classification of folds into concentric, similar, chevron, ptygmatic, and so on, has recently been amplified in a very useful way by Ramsay (1967, pp. 345–372) into what may well become the standard geometric nomenclature for simple types of fold in true profile. This classification permits the shape of a discrete layer or of the part of a uniform rock between two arbitrarily selected folded surfaces to be placed in one of five clearly defined classes. One of these classes (1 B) corresponds to the ideal concentric or parallel fold with constant orthogonal layer thickness, and another class (2) corresponds to the ideal similar fold with constant layer thickness measured in the direction of the axial plane. Other simple types of fold form a continous spectrum with these ideal shapes.

Most of the folds shown in the plates can be classified according to this scheme. In a multilayered sequence folds generally change their class from layer to layer. The great importance of this classification lies in the accurate information it provides on the shapes of folded layers of different composition and thus on the strains involved in the formation of the folds.

No attempt has been made to classify the folds illustrated in the plates*. They do however cover most of the types encountered in routine field work and all of RAMSAY's five classes are represented. There are included also common fold types that do not fit so easily into this scheme—particularly conjugate or box-folds with double hinges, and chevron or kink-folds which have vanishingly small hinge zones and planar limbs. These types of fold can occur in rocks of all kinds from bedded sediments to strongly foliated schists and gneisses.

Folds provide important geometric information to the structural geologist. The trend and plunge of minor folds can be a guide to the orientation of the axes of major folds and nappes. The orientation of the axial planes of minor folds is likewise important and the orientation of the enveloping or median surfaces of trains of adjacent folds can be guides to the direction of stratigraphic continuity ("sheet dip") in strongly folded sedimentary rocks. Most structural geologists are also familiar with the importance of the sense of asymmetry or the "vergence" of minor folds in the structural analysis of regions containing large folds or nappes.

The folds shown in the plates are arranged in three groups. The first group (Plates 67 to 88) includes folds in the bedding of unmetamorphosed or weakly metamorphosed rocks. The second group (Plates 86 to 111) includes folds in metamorphic and generally more strongly deformed rocks. In some of these the folded surface is bedding, in others it is not. Some of the folds shown are associated with axial plane foliations and incipient transposition of the folded surfaces. The third group (Plates 112 to 140) includes folds in strongly foliated rocks, generally phyllites and schists. The folded surfaces are much more pervasive than the layers in folds of the earlier groups and they are of metamorphic origin. Folds in this group tend to be of the kink or

* An excellent exercise for a student is to do this himself by means of transparent overlays placed over the photograph.

chevron type with sharp hinges and planar limbs. Note, however, that folds of this general shape occur also in regularly layered or bedded rocks, but generally on a larger scale (Plates 76 to 78). Conjugate or "box" folds also can occur in a variety of rock types from massive bedded sedimentary rocks (Plate 79) to foliated schists and gneisses (Plates 121, 122 and 140 A).

It is difficult to comment on the origin and significance of folds without taking into account their relationships to the other structures of deformed rocks such as boudins, foliations and lineations. The abundance of folds, in even rocks without these other structures, is clearly traceable to the bedded, layered, or foliated nature of most continental rocks. Bodies with such internal planar structural anisotropies are clearly mechanically unstable when deformed, and folds and boudins appear to be the most obvious expressions of this instability. Folding of most kinds, particularly buckle, flexural slip and kink folding, appears to result from instability of a layered body when compressional stresses act in or close to the plane of the layers. Conversely, boudinage, pinch and swell structure (necking) and extension fracturing (see below) appear to result from instability where the same direction is one of extension. Other types of folds have been recognized. Bending folds produced, for example, by differential uplift of an underlying basement are formed by forces transverse to the folded layers and may not indicate layer-parallel shortening. Slip folds, believed by some geologists to result from some material instability in a flowing rock, likewise may have different kinematic and dynamic significance. But most folds seem to result from the geometric instability (shape instability) of bodies with planar heterogeneity or anisotropy and in principle to be structures resulting from buckling of various kinds (including kinking). As such they are indicators of the directions and magnitudes of regional strains and displacements.

The possible profile shapes of common minor folds are well represented in the plates. These shapes depend closely upon the nature of the layering or foliation. In general, folds in massively bedded or layered rocks are larger (Plate 82) than in rocks of similar kind with thin beds or layers (Plate 76 A). In most bodies folds of several orders of size are present (Plate 90 A), although this fact is not always obvious from the minor structures alone. Folds of broadly concentric style tend to occur in coarsely layered sequences (Plate 98 A) or in thick "competent" layers enclosed in a less competent matrix (Plate 89). Concentrically folded layers can occur also in the more competent members of rhythmically layered sequences (Plate 81). Folds of similar style occur most commonly in metamorphosed rocks—particularly in marbles (Plates 102 and 111 A), but also in quartzites and feldspathic rocks (Plates 101 and 104). Folds of the chevron or kink type are most typical of well foliated or finely bedded rocks. Some of these folds form parallel-sided shear domains in otherwise unfolded rocks and are generally termed "kink bands" (Plates 114 to 120). Such structures grade either by conjugate kinking (Plates 121 to 123) or by increase in number and tightness of the kink bands into chevron folds of various kinds (Plates 127 to 140).

Folding is generally the precursor of transposition of one planar structure into another. Tight folding is generally accompanied by the development of axial plane foliation or by attenuation and disruption of fold limbs (Plates 13, 20, 21, 26 and 100 A). Complete disruption of folded layers, where transposition is far advanced, leads initially to the formation of rootless or intrafolial folds (fold boudins)—tightly folded hinges separated from their attenuated limbs and floating in a foliated matrix (Plates 19 B, 21 B, 42 and 107). Ultimately these fragments too can become so flattened in the plane of the foliation that their true nature is lost and all trace of the episode of folding disappears.

Boudins (Plates 141 to 155)

In comparison with folds, boudins seem to be relatively rare structures, and have consequently received less attention from geologists. Boudinage appears to be an expression of layer instability in extension, and it occurs, in the simple form illustrated in most textbooks, only in layers that have been progressively extended by small to moderate amounts. Most boudin formation is preceded by a phase of folding so that intrafolial folds or fold boudins are far commoner than the classical types of barrel or elliptically shaped boudins.

Structures closely related to boudins also formed in progressively extended layers are extension fractures (commonly vein-filled) and "pinch and swell" structures or "necks." Both structures can be viewed as precursors to boudin formation and they illustrate, as do the profile shapes of boudins, the effects of both the material property contrasts between a layer and the surrounding material, and the physical conditions under which the boudinage occurred. If a layer is relatively strong and brittle in comparison to the enclosing matrix, it tends to fail by fracture, either in extension or shear, and separates into blocky, angular boudins (Plate 141 A). As the boudins drift apart the more ductile matrix can flow into the spaces which would otherwise be left unsupported (Plates 142 A, 143 A and 144 A). Between some boudins vein material segregates to fill these spaces. Other boudins appear to have formed in layers with ductility more similar to that of the enclosing material. These acquire a more elliptical or spindle-shaped profile as a result of separation by necking rather than fracture (Plates 145 to 148). Other types are intermediate in shape (Plates 141 B and 142 B) and some have the classic barrel-shaped form generally illustrated in textbooks (Plate 144B). Extreme necking can result in "beaded" rows of boudins with thread-like connections (Plate 150).

In three dimensions, many boudins are rod-like in shape: separation seems to occur at right angles to the extension direction. But some boudins are ellipsoidal and have clearly resulted from extensions in all directions in the plane of the foliation. Because boudins are usually difficult to observe in three dimensions in the field, this phenomenon may be more widespread than is generally believed.

In some regions, veins or dikes oblique to a principal layering may also be folded or boudinaged (Plates 149, 170, 173 and 174). Careful study of the effects of progressive shortening or extension on layers with different initial orientations in a single body promises to yield the most complete information on regional strain patterns, and also to permit the interpretation of associated structures such as foliations and lineations in terms of measurable strains.

Boudinage, like folding, is intimately bound up with progressive transposition of planar structures; and, as with folding, the greatest attention has been given either to the behavior of isolated layers of one material in a matrix of another, or to multilayers of different materials. The instability of such composite materials in compression and extension is easily pictured. But recent studies of chevron and kink folds both in the field and by experiment have thrown light on the folding instability of pervasively foliated homogeneous materials. No such attention has yet been given to the behavior of the same kind of materials during progressive extension. There must exist in rocks of this kind an analogue of boudinage of heterogeneously layered materials, unless, as seems unlikely, uniformly foliated materials are always mechanically stable and extend homogeneously without local instability. More probably, the instability is either on the microscopic scale, or a "phacoidal" or "braided" foliation develops such perhaps as that illustrated in Plate 36. Thus a regular foliation may be the analogue in a homogeneous rock of boudinage and *mélange* formation in grossly extended heterogeneously layered rocks.

Fractures and Veins (Plates 156 to 174)

A feature of many deformed rocks is the presence of vein-filled fracture systems of various kinds. The vein materials can be a variety of minerals but are commonly quartz or calcite. Such fracture systems are of interest to the structural geologist for two main reasons. First, the fractures themselves are a result of deformation; and some kind of strain analysis, or even stress analysis. if the aggregate strains are very small, may be possible. Second, such vein-filled fractures can develop at various stages during progressive deformation, and they can thus play the role of planar markers implanted in a body at various stages along a deformation path. The subsequent distortion of such markers emplaced at different times in different directions can yield information about both the state of strain and the path by which it has been reached.

Most fracture systems in rocks are interpreted either as *extension fractures*—formed more or less at right angles to a principal extension, or as *shear fractures*—obliquely inclined to principal strain directions, commonly in conjugate sets. Unequivocal separation of these two types in the field is commonly difficult and many types observed in nature appear to be intermediate between these ideals.

Examples of what appear to be quartz vein-filled extension fractures developed at right angles to competent beds of graywacke are given in Plates 156 to 160. The fractures are in parallel sets and the best examples (Plate 158) are clearly associated with "necks" and pinch and swell structure in the beds. In another rock with beds of greywacke more widely separated in a less competent material such as shale, boudinage would probably have developed by the stage in deformation shown in this plate. Some veins are wedge-shaped, being much wider on one side of the bed than the other (Plate 158). Others do not cross a bed completely (Plate 156), and still others are strongly elliptical (Plate 160) being thickest at the center of the bed and almost dying out at the margins. Veins of this shape appear to fill fractures that have propagated from the center of a bed towards its margins.

In some folded rocks, extension fractures, pinch and swell structures and a general layer thinning are present on the limbs of the folds. In others these structures can be confined to one limb (Plate 159), suggesting that, even in symmetric folds, the strain histories on opposite limbs need not be the same.

The *en échelon* pattern of vein-filled fractures (Plates 162 to 168) is one of the commonest in homogeneous rocks. Some of the individual veins are elliptical in shape (Plate 165); others are sigmoidally curved (Plate 164). In some beds or layers *en échelon* veins are associated with extension fractures (Plate 162) and in favorable circumstances conjugate sets of *en échelon* veins are developed (Plates 166 and 167). In extreme instances vein material can occur in a complex network of fractures of many kinds and constitute a significant proportion of the rock (Plate 168).

The importance of distorted veins in regional strain analysis has been commented upon in the last section. Examples of veins distorted by later movements are given in Plates 161 and 169 to 174. Veins can be folded, boudinaged or both (Plate 172). Displacement of veins can confirm interlayer slip in layered or bedded rocks (Plate 161).

Deformed Primary Structures (Plates 175 to 186)

The most useful indicators of local strain in deformed rocks are distorted primary structures of known or determinable initial form. A great variety of such structures have been used for analysis of the state of finite strain some of which are illustrated in the plates. Examples are, pebbles (Plates 175 to 179), primary sedimentary structures (Plates 180 to 182), and fossils (Plates 183 to 186). Other structures that can be used include ooliths, pisolites, spherulites, reduction spots and certain igneous structures such as pillows (in lavas) and amygdules.

Complex Structures (Plates 187 to 203)

The final group of plates in this collection illustrates some of the structural complexities found in rocks which have been deformed in more than one episode. A permanent deformation of sufficient magnitude results in the formation of structures of the kind described above. In rocks affected by more than one period of deformation, where the last was insufficiently intense to have obliterated all earlier structures, generally distinct generations of structures can be present, each generation overprinting and modifying all earlier generations. Even the simplest rocks of this kind tend to contain more than one foliation and lineation, and most of them contain also two or more sets of minor folds which interfere to form patterns of bewildering complexity.

An understanding of such complex structures can come only by way of a full appreciation of the properties of the simpler minor structures illustrated in the foregoing plates. The study of complex structures remains today very difficult and time-consuming because of the large number of data which must generally be gathered in the field and the special techniques required to interpret them.

The most that can be done here is to provide illustrations of some of the commoner structural indications of polyphase deformation in rock bodies, such as: intersecting lineations of different ages (Plates 187 to 189); earlier lineations bent by later folding (Plates 190 to 193); earlier folds crossed by a later foliation (Plates 194, 195 and 201 A); interference of two generations of folds of similar amplitude (Plates 196 to 203).

The possible relationships between superposed generations of deformation structures are very variable and *a priori* unpredictable. In no branch of the field study of deformed rocks is the ingenuity and imagination of the field geologist taxed so heavily to arrive at a satisfactory interpretation. Each new generation of structures modifies or obliterates the older ones, generally in a patchy and irregular way, and what fragmentary evidence survives must be carefully examined, correlated and interpreted in as reasonable a fashion as possible. Many carefully conducted studies, some of which are listed in the bibliography, have shown that a coherent picture of successive episodes of regional deformation can emerge even for a large and complex body of deformed rock, such as the Scottish Highlands. Present interest in gross crustal motions, as pictured in the plate tectonic theory, provides a new impetus for the study of complexly deformed bodies, and offers the hope that interpretations of presently local significance may one day form part of a coherent picture of past large-scale motions in the crust.

1 A. Cleavage-fractures or "fissures" cross a layer of massive quartzite. The fractures are inclined obliquely to slaty cleavage in adjacent shales (lower left). Pre-Cambrian, Porsanger Fjord, Finnmark, Norway

1 B. Cleavage fractures in folded layer of quartzite are inclined symmetrically to the axial plane of the fold. Dalradian, near Whitehills, Banffshire, Scotland

2. Cleavage fractures are associated with open folding in beds of graywacke. The fractures are more closely spaced in fine-grained layers. Material from these layers penetrates locally into widely spaced fractures in the more massive beds. Paleozoic, near Bateman's Bay, New South Wales, Australia

3 A. Discrete cleavage fractures are present in a single bed of graywacke enclosed in pervasively cleaved mudstones. Fractures in the graywacke are inclined obliquely to the slaty cleavage. Lower Devonian, Col d'Aspin, Pyrenées, France

3 B. Folding and cleavage are developed in bedded sandstones and siltstones. In a coarse-grained layer (upper center) discrete cleavage fractures inclined symmetrically to the axial plane of a fold are present. In fine-grained rocks a more pervasive cleavage is subparallel to the axial plane of the fold. Permo-Carboniferous, near Matesevo, Montenegro, Yugoslavia

4 A. Graywacke beds dip steeply left and are crossed by cleavage parallel to the hammer handle. The cleavage is weak but pervasive and the discrete fractures have slight sigmoidal curvature across each bed. Pre-Cambrian, South Stack, Holy Island, Wales

4 B. Bedded graywackes and siltstones dip steeply left. The character of the cleavage in each layer reflects its composition and grain size. A coarse graywacke (right) has discrete subhorizontal cleavage fractures; a medium-grained siltstone (left, beneath hammer) has moderately dipping pervasive cleavage. The most steeply dipping and finely pervasive cleavage (center) is in the layer with finest grain. From the same locality as Plate 4 A

5. Structural features are similar to those of Plate 4 B. From the same locality as Plates 4 A and 4 B

6 A. Folded graywackes and siltstones contain cleavage and fractures subparallel to the axial planes of the folds. From the same locality as Plates 4 A, 4 B and 5 (height of exposure about 50 m)

6 B. In massive graywacke layers, tight small-scale folding and well developed cleavage is present. Incipient disruption and transposition of thinner beds is evident. Detail from Plate 6 A (height of exposure about 15 m)

7A. Subhorizontal bedding in graywacke is crossed by a coarsely developed prominent cleavage, dipping steeply left. The cleavage is parallel to the axial plane of a gentle fold in the bedding. Dalradian, Rosehearty, Aberdeenshire, Scotland

7B. Pervasively cleaved mudstones show no obvious bedding. Lower Devonian, Col d'Aspin, Pyrenées, France

8. Along the margin of a body of spilitic volcanic rocks enclosed in metamorphosed sedimentary rocks a platy foliation (dipping steeply left) is developed. Pre-Cambrian, Llanddwyn Island, Anglesey, Wales

9A. A finely pervasive slaty cleavage (dipping gently right) is developed in silty beds (horizontal) in otherwise uncleaved graywackes. Silurian, near Kirkudbright, Scotland

9B. Subhorizontal bedding in mudstones and siltstones is crossed by finely pervasive cleavage (dipping steeply right). Some layers are more strongly cleaved than others. Lower Devonian, Col d'Aspin, Pyrenées, France

Estwing

10. Bedded graywackes and siltstones dip moderately to the right and contain a more gently dipping pervasive cleavage. Dalradian, near Dunoon, Argyllshire, Scotland (height of exposure about 10 m)

11 A. Finely developed bedding laminations in tillite dip gently right and are crossed by a regular pervasive cleavage parallel to the hammer handle. Large fragments of crystalline rock (such as granite) in the tillite are apparently unaffected by the cleavage. Pre-Cambrian, near Tana, Finnmark, Norway

11 B. Slaty cleavage (parallel to hammer handle) is parallel to the axial planes of gentle folds in faint bedding laminations, dipping left. From the same locality as Plate 11 A

12. Planar bedding laminations in calcareous mudstones dip to the right and are crossed by more steeply dipping regular slaty cleavage. High Calcareous Alps, Switzerland

13 A. Prominent cleavage in both coarse and fine beds of graywacke is inclined symmetrically to axial planes of folds. Incipient transposition of bedding is evident in some layers. Pre-Cambrian, near South Stack, Holy Island, Wales

13 B. Cleavage subparallel to the axial plane of a fold in bedded graywackes and siltstones is most pervasively developed in the darker, less sandy layers. A small fault displaces the upper limb of the fold. Carbonifeorus, Millook Haven, Devonshire, England

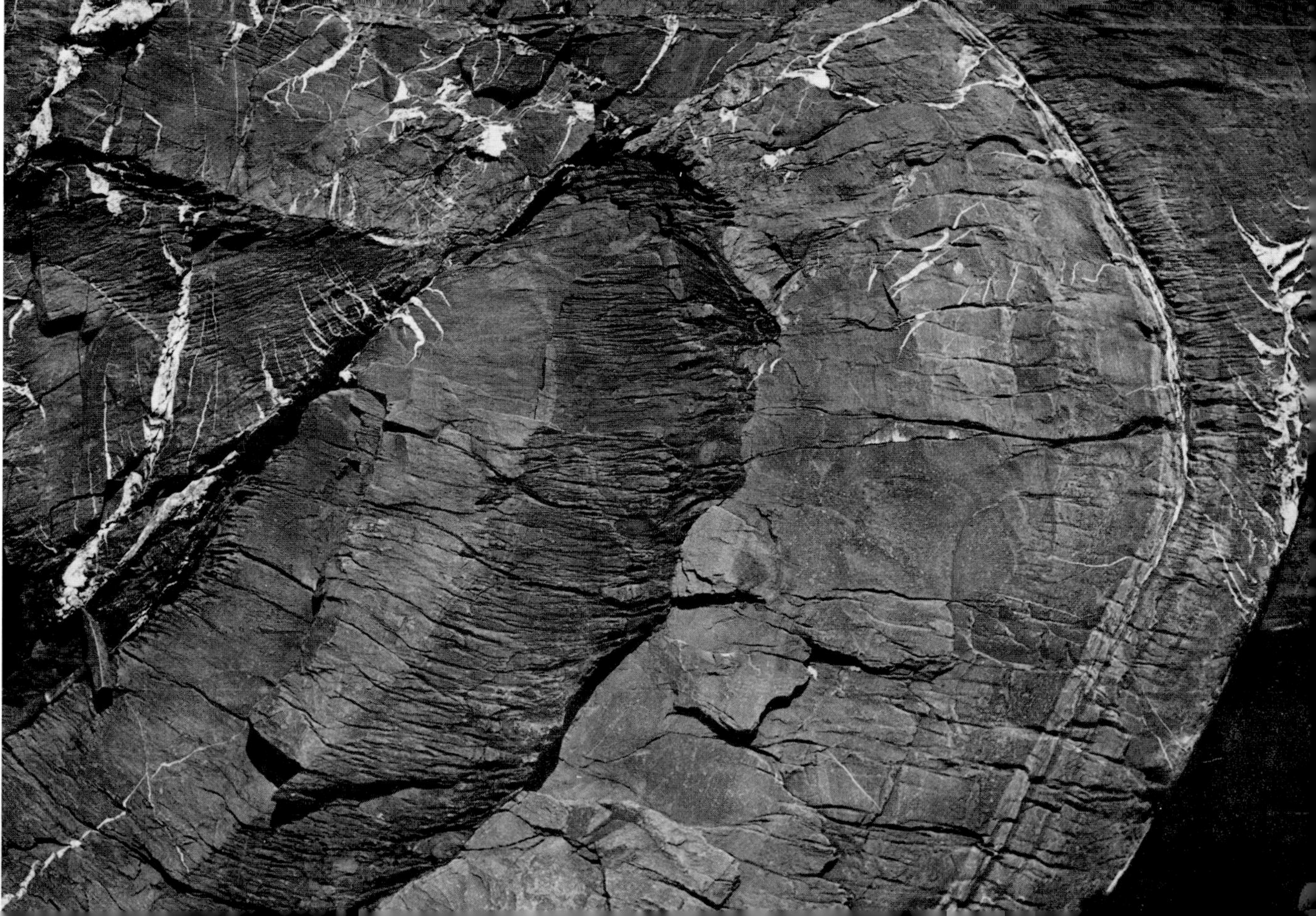

14. Detail of Plate 13 B

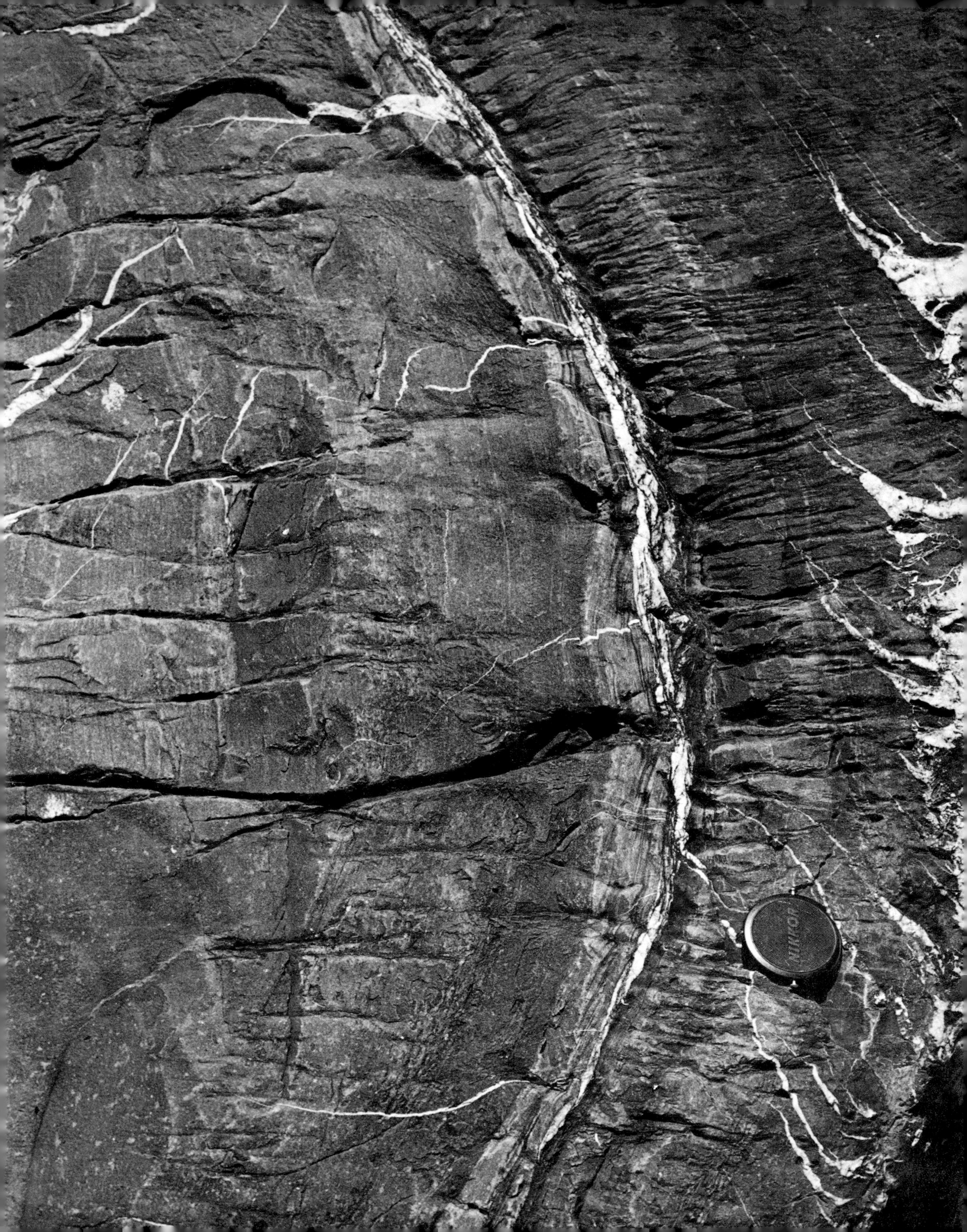

15. Axial plane cleavage is most strongly developed in fine-grained bed in core of the fold. From the same locality as Plates 13 B and 14

16. A prominent slaty cleavage dips steeply left across more gently dipping bedding in slate. Planar quartz veins are locally developed along cleavage surfaces (lower left). Devonian, Ilfracombe, Devonshire, England

17. A well developed cleavage dipping steeply left crosses relict bedding in phyllitic slate, dipping gently left. In more coarse-grained beds cleavage is less pervasive and sigmoidally curved. From the same locality as Plate 16

18A. In an otherwise uniformly foliated phyllitic slate a single bed of coarser material is preserved dipping gently left. From the same locality as Plates 16 and 17

18B. The regular pervasive slaty cleavage dipping steeply left is typical of commercial slates. No bedding is visible. Cambrian, near Llanberis, Caernarvonshire, Wales

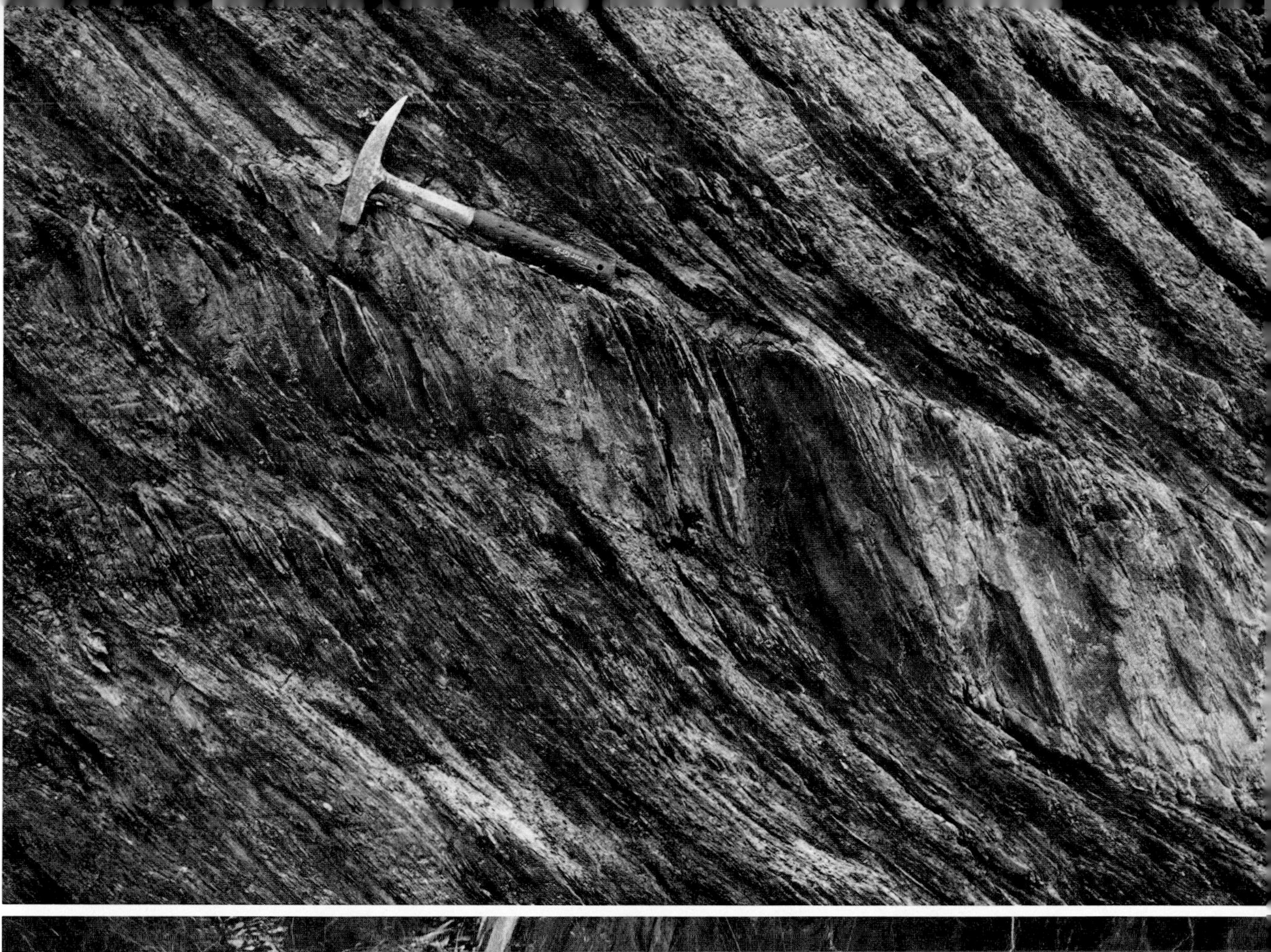

19 A. Cleavages of different types are present in interbedded graywackes and siltstones. Pre-Cambrian, South Stack, Holy Island, Wales

19 B. Detail of Plate 19 A showing advanced local transposition of bedding laminae in one layer. Attenuated limbs of tight folds define a "foliation" which passes longitudinally into discrete cleavage fractures in coarser beds

Ever Grip

20 A. Folding of phyllite is associated with a secondary foliation subparallel to the axial planes of the folds. Pre-Cambrian, Tre-Arddur Bay, Holy Island, Wales

20 B. Detail of Plate 20 A showing that the secondary foliation (dipping steeply left) is defined by the attenuated limbs of small crenulations in the phyllitic foliation

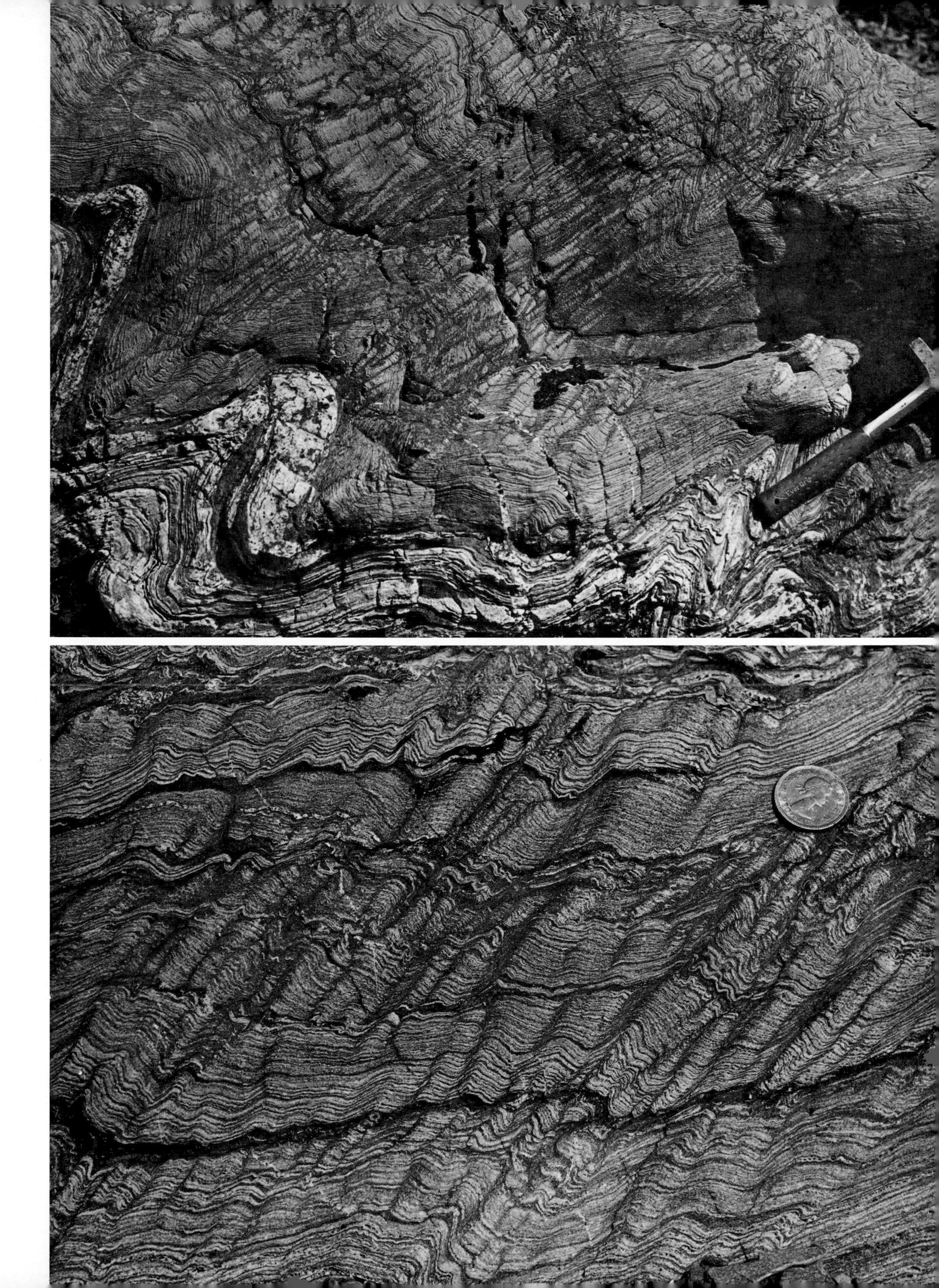

21 A. Bedding in graywacke and siltstone is tightly folded and disrupted during the development of a horizontal transposition foliation. Carbonifeorus, near Brest, Brittany, France

21 B. The foliation in a schist is folded and crossed by a secondary foliation dipping steeply right. Paleozoic, La Viella, Pyrenées, Andorra

22 A. A network of thin quartz veins in mica schist is folded and incipiently transposed during development of a horizontal foliation. Paleozoic, near Cooma, New South Wales, Australia

22 B. In the hinge of a fold relict beds of quartzite (subvertical) are at right angles to foliation in enclosing mica schist. From the same locality as Plate 22 A

23. Layers and lenticles of quartzite are transposed remnants of sedimentary beds in a foliated mica schist. The layers and the foliation are folded and a secondary foliation of the "strain-slip" or "crenulation" type is developed parallel to the axial plane of the fold. Dalradian, Loch Leven, Inverness-shire, Scotland

24A. A horizontal mica-rich layer in quartzite contains an incipient foliation (parallel to pencil) defined by attenuated limbs of weak crenulations. Mica is concentrated in these surfaces. Dalradian, Loch Leven, Inverness-shire, Scotland

24B. In quartz-mica schist an earlier transposition foliation (dipping left) is crossed by a later crenulation foliation (parallel to pencil). From the same locality as Plate 24A

VENUS
MADE IN ENGLAND - VENUS PENCIL CO. LTD.

"KING'S OWN"
HB

25. In quartz-mica schist an earlier foliation defined by alternating quartz and mica-rich laminae is crossed by a later coarse foliation of the strain-slip or crenulation type. The "surfaces" of this later foliation are discretely spaced layers in which the earlier laminae are attenuated. From the same locality as Plates 24A and 24B

MADE IN ENGLAND
VENUS

26 A. A later vertical foliation in mica schist is parallel to the axial planes of folds in an earlier foliation. Mica is concentrated along the later foliation "surfaces" which are actually thin zones of laminar displacement, as can be seen from offsets in the light-colored quartz-rich layers and lenticles parallel to the earlier foliation. Paleozoic, Sutter Creek, California, U.S.A.

26 B. Structural features resemble those of Plate 26 A but transposition of the earlier into the later foliation is more advanced. A thicker quartzitic layer (left of center) is internally folded and locally displaced upon relatively few discrete foliation surfaces which are slightly "refracted" at the boundary of the layer. From the same locality as Plate 26 A

27. Structural features resemble those of Plates 26A and 26B. Details of folding and incipient transposition of the earlier foliation by the later (vertical) foliation can be clearly seen. Thin dark laminae rich in mica are present along the later foliation. From the same locality as Plates 26A, 26B and 27

28. Structural features are similar to those of Plates 26A, 26B and 27

29A. A very regular foliation is present in strongly deformed quartzitic Valdres Sparagmite. All primary structures are transposed and the uniform laminae of tectonic origin bear a superficial resemblance to sedimentary bedding. Pre-Cambrian, near Bygdin, Oppland, Norway

29B. Detail of Plate 29A

30A. The rock is an intimate mixture (*mélange*) of wavily foliated phyllite and quartz lenses, some of which may have originated as veins or similar segregations. All earlier structures are transposed. Lower Paleozoic, near Gjendesheim, Oppland, Norway

30B. In this *mélange* traces of isoclinal folds are preserved in some of the quartz-lenticles, but transposition of earlier structures is virtually complete. The structure is probably a more advanced stage of that in Plate 27. Lower Paleozoic, near Randsverk, Oppland, Norway

ESTWING

31 A. No undoubted bedding survives in a *mélange* of quartzite lenses in graywacke. The lenses may be disrupted fragments of initially planar beds. Dalradian, near Whitehills, Banffshire, Scotland

31 B. The strongly flattened quartz lenses in mica schist may in part represent disrupted quartz veins and rods. The extensive attenuation of these fragments suggests local deformation of large magnitude. Bedding and other primary structures, if once present, have been completely destroyed. Moine, Kyle of Tongue, Sutherland, Scotland

32A. This *mélange* of quartz lenses in greenschist is of interest as the first to be so named (the Gwna *mélange*, Greenly, 1919). All local primary structures are destroyed. The foliation of the *mélange* is commonly tightly folded. Pre-Cambrian Mona Complex, Menai Bridge, Anglesey, Wales

32B. In a gneissic *mélange* quartzo-feldspathic lenses and "pods" float in a micaceous matrix. The gneiss may represent a sheared plutonic rock. St. Malo, Brittany, France

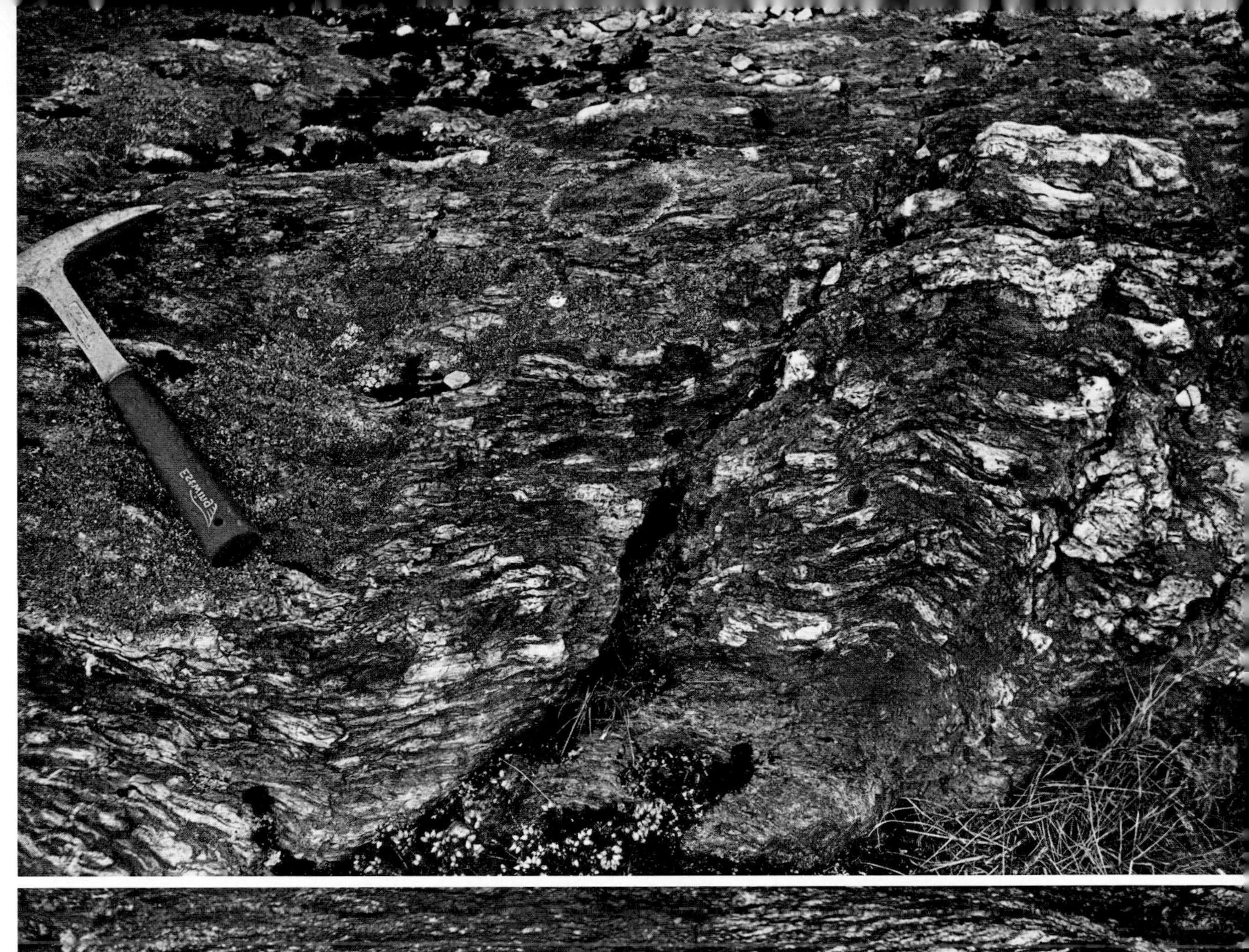
Estwing

Estwing

33 A. Quartz lenses, possibly disrupted veins, are flattened in the plane of a regular foliation in mica schist. Le Conquet, Brittany, France

33 B. Massive quartzo-feldspathic lenses float in a foliated gneissic matrix. Ste. Maxime, Cot d'Azur, France

34. Foliation in a phyllitic mica schist is wavy but uniform. No trace of bedding or other primary structures is preserved. Paleozoic, near Amsteg, Uri, Switzerland

35 A. In strongly foliated mica schist small elliptical bodies of quartz are present flattened in the plane of the foliation. Paleozoic, Andermatt, Uri, Switzerland

35 B. No bedding is preserved in strongly foliated *mélange* of *schistes lustrés*. Mesozoic or Tertiary, near Bessans, French Alps

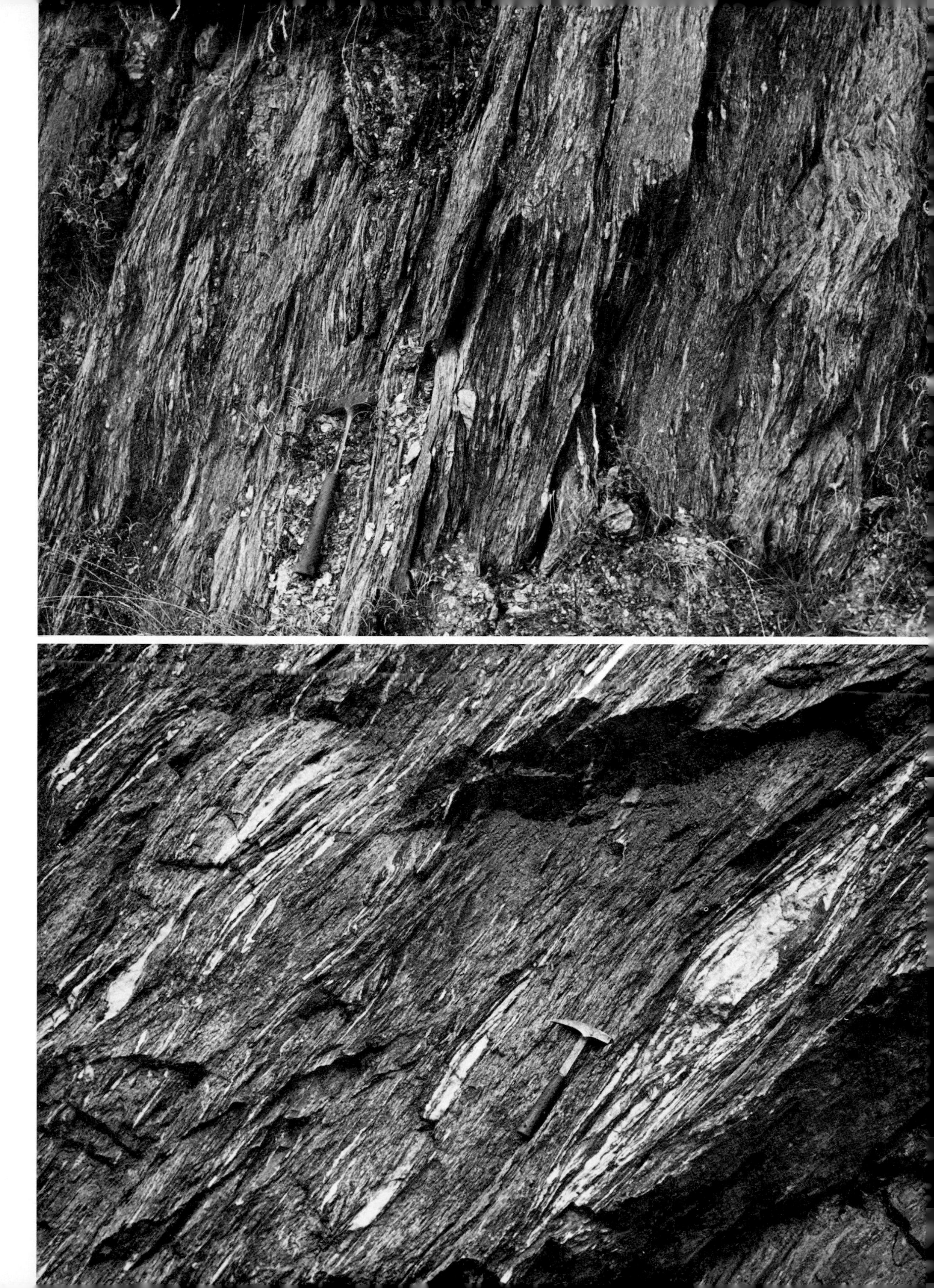

36. Foliation in a quartz-mica schist is defined by overlapping lenticular aggregates of quartz separated by films of mica and other minerals. Some of the lenticular fragments resemble flattened pebbles or fragments of layers. Oberwald, Valais, Switzerland

37. A coarse lenticular foliation in gneiss is defined by quartzo-feldspathic and micaceous aggregates. Near Neufenen, Valais, Switzerland

Ever Grip

38 A. A very regular platy foliation is present in a quartz-schist. No primary structures survive. Near Volos, Greece

38 B. Lenticular aggregates of quartz form a *mélange* with strongly foliated mica schist. Saastal, Valais, Switzerland

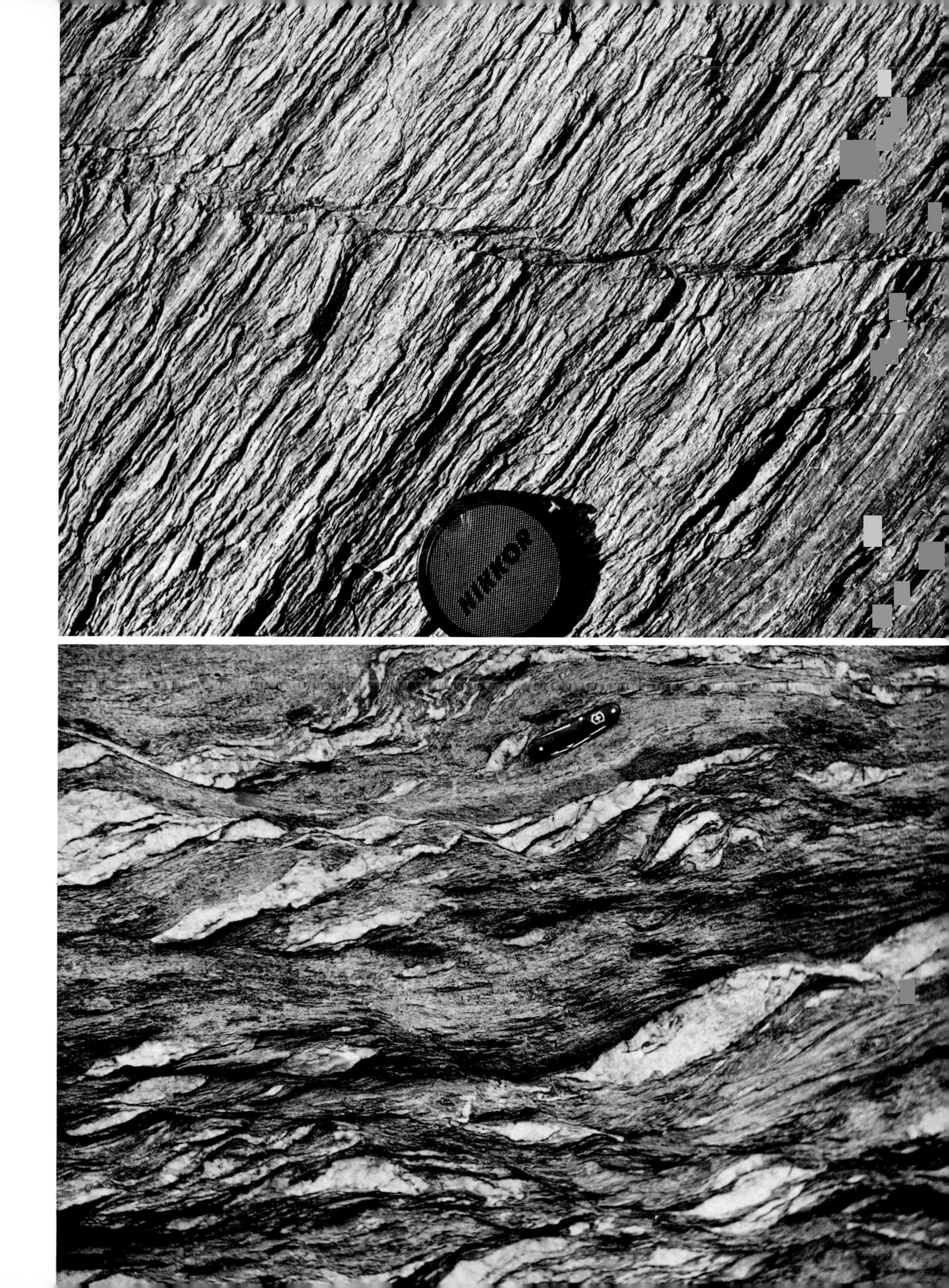
NIKKOR

39A. Vertically foliated mica schists are in the core of a large plunging fold (not shown). The foliation is parallel to the axial plane of the fold and bedding is for the most part transposed. Near Weldon, Sierra Nevada, California, U.S.A.

39B. This "Franciscan" type *mélange* is an intimate mixture of disrupted fragments of chert, graywacke and basalt in a schistose matrix. All primary continuity of layers is locally destroyed and intermixing of rock types has occurred on a very small scale. Cedros Island, Baja California, Mexico

40. In the same *mélange* as shown in Plate 39 B a larger ellipsoidal body of graywacke "floats" in a schistose matrix

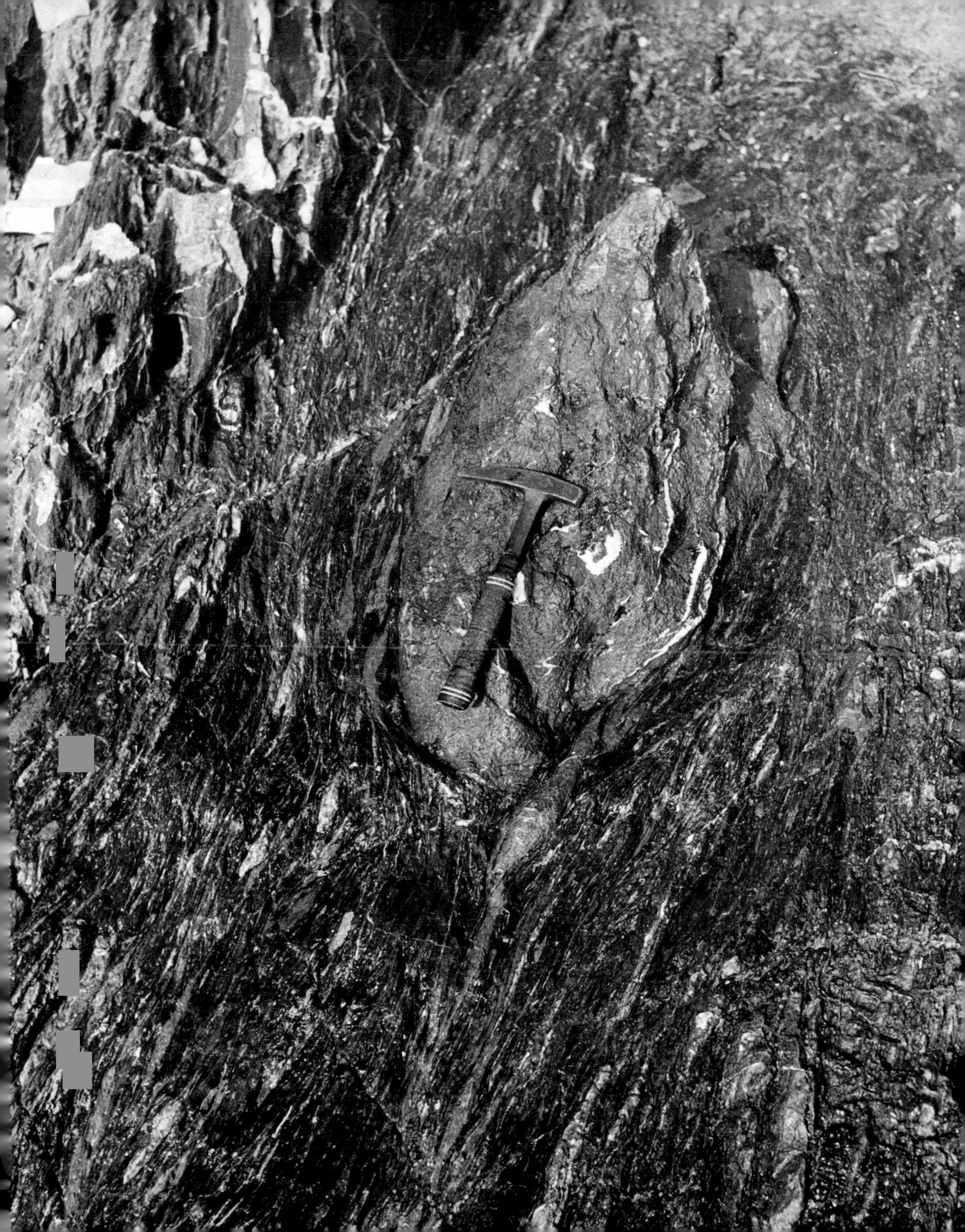

41. No bedding survives in this very regularly foliated quartz-mica schist. Above the hammer handle can be seen an isoclinal fold in a very thin quartz vein. Dalradian, Loch Leven, Inverness-shire, Scotland

42. In an otherwise uniformly foliated phyllite, a "rootless" fold in a quartzite layer survives. Such relict fold hinges generally indicate that transposition of earlier layering has occurred. Lower Paleozoic, near Gjendesheim, Oppland, Norway

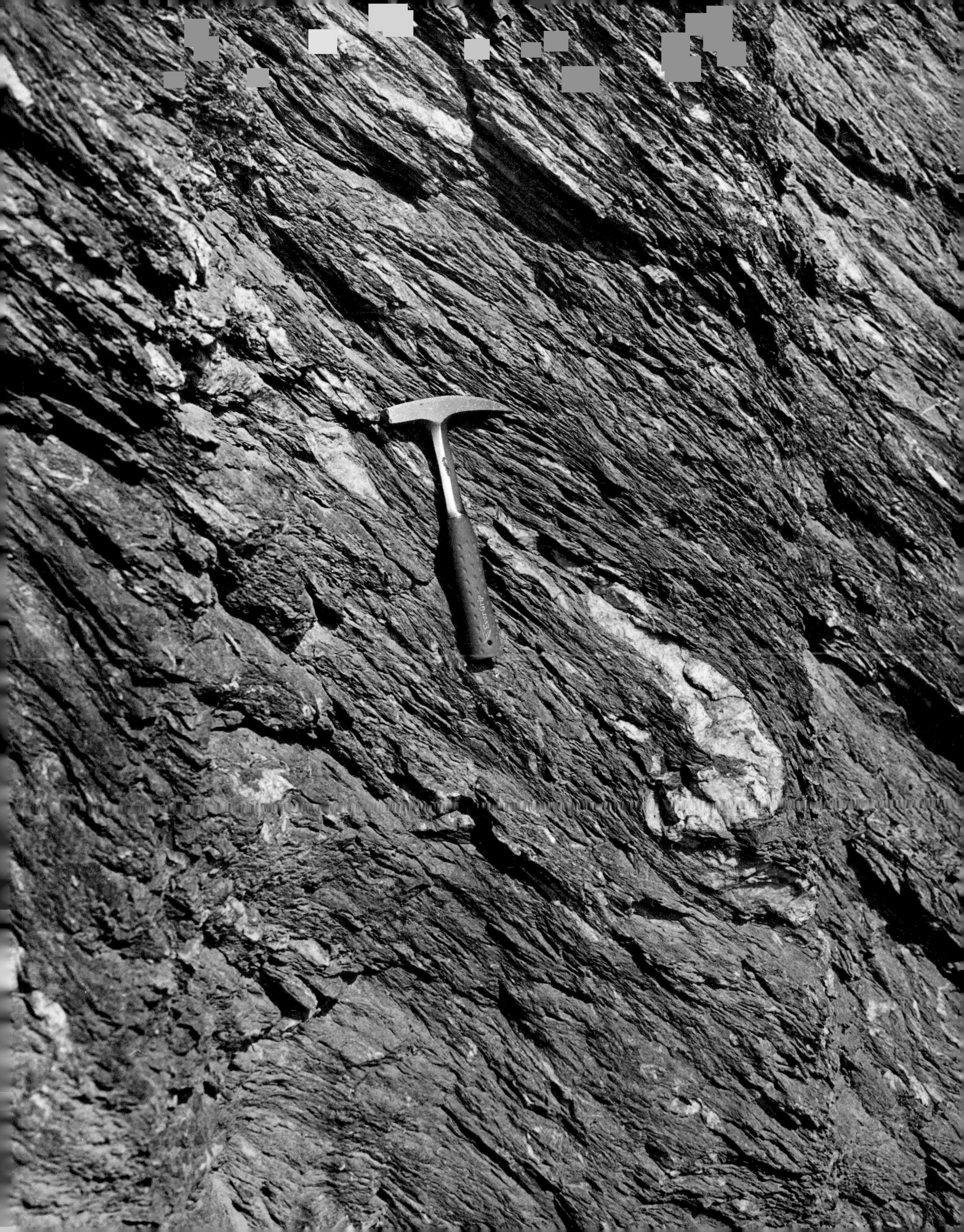
Ever Grip

43. In an "augen" gneiss large rounded crystals and aggregates of feldspar float in a conspicuously foliated matrix. Paleozoic, Oppdal, Sör-Tröndelag, Norway

44 A. Detail of Plate 43

44 B. In a foliated and lineated gneiss large feldspar crystals are prismatic in shape and have a linear preferred orientation. Ste. Maxime, Cot d'Azur, France

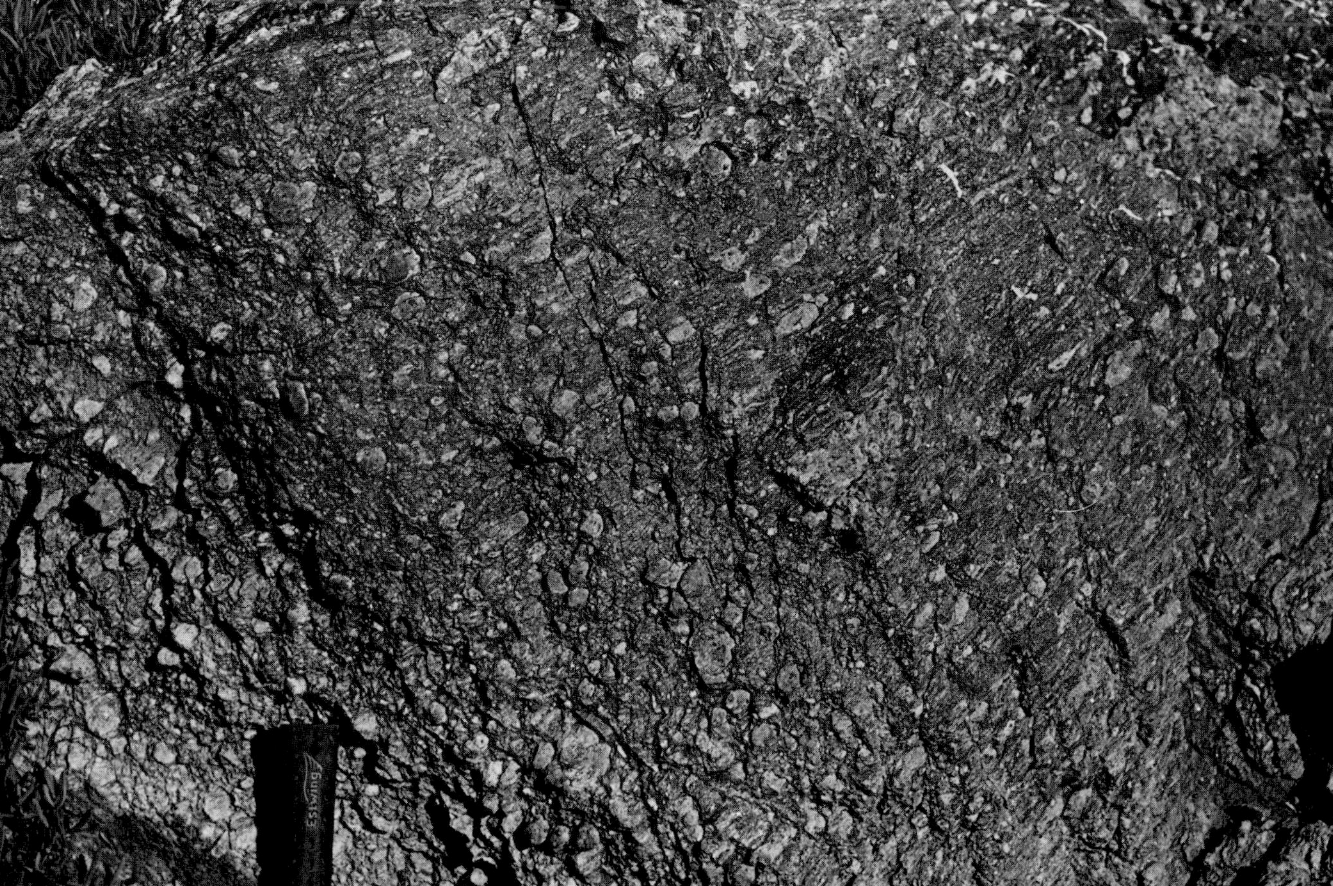

45. In quartzitic Valdres Sparagmite intense deformation has formed a very regular platy foliation. No undoubted primary structures survive. Near Bygdin, Oppland, Norway

46

46 A. The lenticular form of many of the layers in a gneiss strongly suggests that the banding does not represent undisturbed sedimentary bedding. Lower Paleozoic, near Mo-i-Rana, Norway

46 B. Such layering and lamination in a flaggy gneiss is commonly termed "bedding foliation," because of the bedded appearance of the rock. It is usually very difficult to establish how much of the banding represents a survival of relatively undisturbed primary structure. Near Karvik, Troms, Norway

47A. The rocks are similar to those shown in Plate 45. A regular planar structure is present in flaggy quartzites, but the laminations do not seem to be of strictly sedimentary origin. Oppdal, Sör-Tröndelag, Norway

47B. In flaggy laminated quartz schists unusually regular foliation and lineation is developed. The lineation appears as fine step-like horizontal striations lying in the foliation surfaces. Pre-Cambrian, near Alta, Finnmark, Norway

48 A. Individual layers in banded gneissic schist are markedly continuous and could represent original sedimentary beds grossly attenuated but not disrupted by deformation. Moine, Lochailort, Inverness-shire, Scotland

48 B. Rocks similar to those of Plate 48 are more gneissic but exhibit markedly continuous layering coupled with an unusually irregular and characteristic pattern of repetition. Such features suggest a sedimentary origin for the layers. Moine, Kinlochmoidart, Inverness-shire, Scotland

49 A. Such lenticular banding is typical of many high-grade gneisses. Some of the lens-shaped aggregates of minerals resemble boudins and may have a similar origin. Pre-Cambrian, Kragerö, Telemark, Norway

49 B. In quartzo-feldspathic gneisses foliation and lineation are strongly developed. In the surrounding region these structures have extremely uniform orientation. Near Turoka, Southern Kenya

Ever Grip

50 A. The light-colored layers are quartzo-feldspathic veins conformable to a uniform regional foliation in mica schist. Le Conquet, Brittany, France

50 B. In a uniformly foliated schist quartz veins are both parallel and obliquely inclined to foliation. Isoclinal folds and boudins are locally present in the veins. Paleozoic, Oberwald, Valais, Switzerland

51. Calc-silicate layers representing original sedimentary beds are preserved in marble. Near Weldon, Sierra Nevada, California, U.S.A.

52 A. Discontinuous crenulations on the foliation surface of a flaggy quartz-mica schist define a very regular lineation. Probably from the Oppdal region, Vågå churchyard, Oppland, Norway

52 B. A coarser crenulation lineation (horizontal) in a more micaceous layer is crossed by weaker crenulations plunging steeply left. From the same locality as Plate 52 A

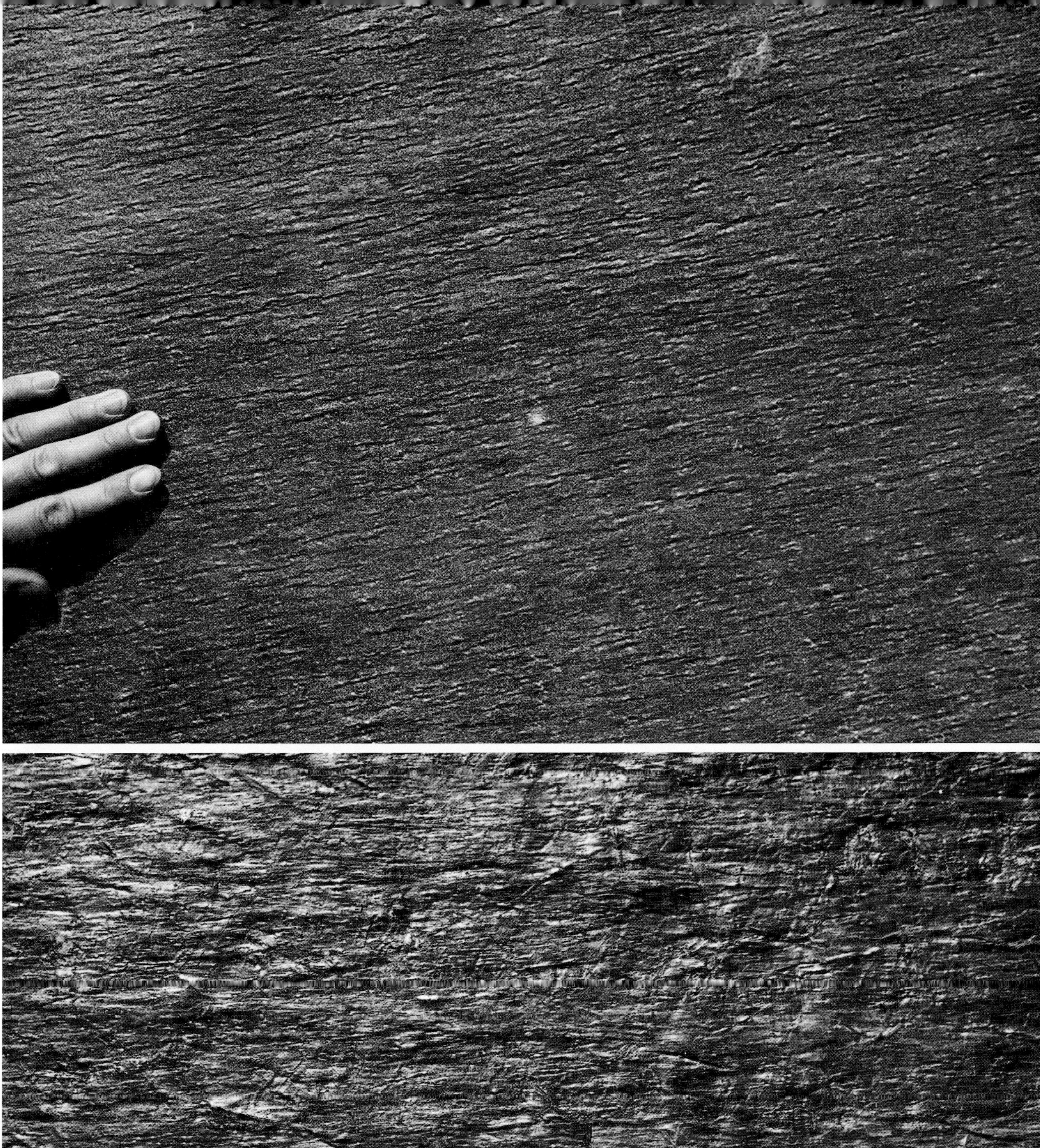
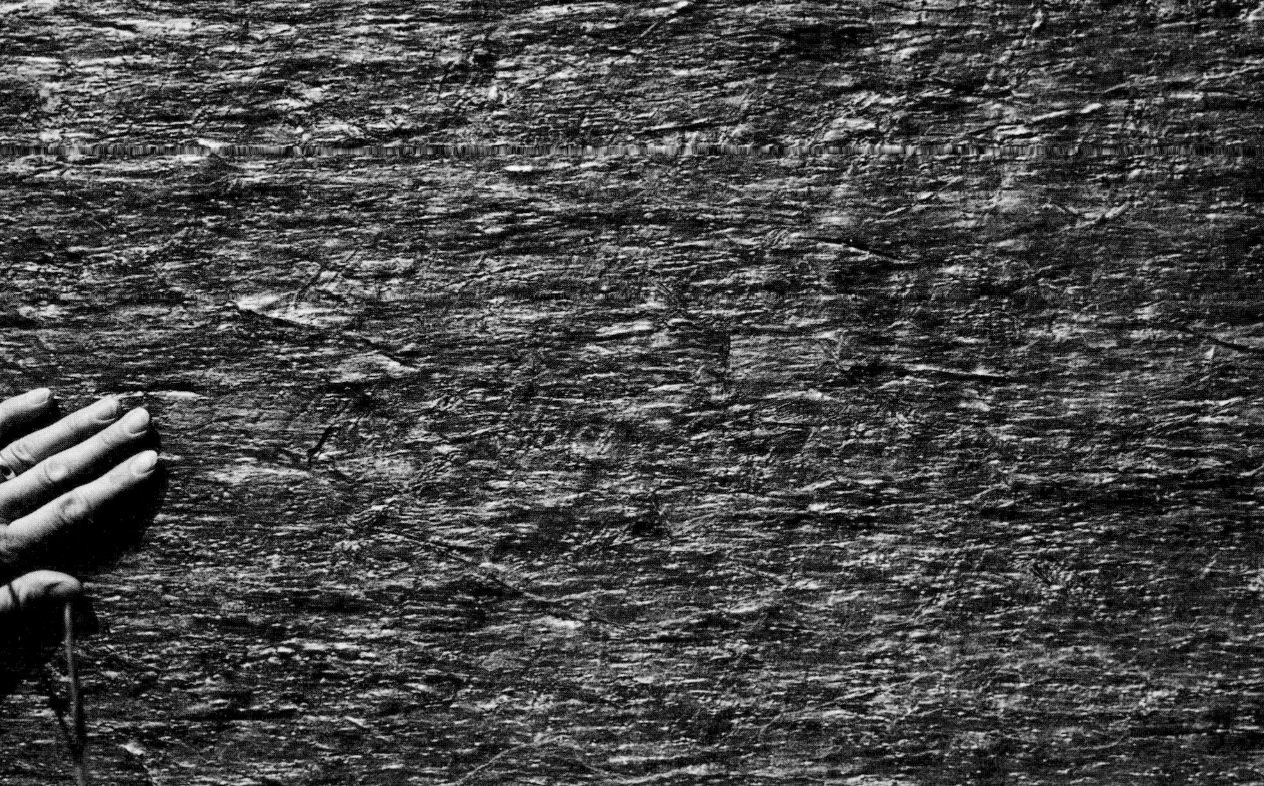

53A. A crenulation lineation plunging right is accompanied by "streaking." From the same locality as Plates 52A and 52B

53B. A prominent crenulation lineation is parallel to the axes of small folds in quartz-mica schist. Lukmanier region, Ticino, Switzerland

54. A coarse crenulation lineation in mica schist plunges right. The fainter more steeply plunging streaks are an older pre-crenulation lineation. From the same locality as Plate 53 B

55 A. A coarse and very regular lineation in schist is defined both by crenulations and by inequant minerals in a strong state of preferred orientation. Small folds with axes in the same direction accompany the lineation. Moine, Glenshiel, Ross and Cromarty, Scotland

55 B. In the same schist a coarse ribbing or mullion structure is defined by the lineation and hinges of small folds. From the same locality as Plate 55 A

56A. Light-colored quartz layers in mica schist are tightly folded on a small scale. A prominent lineation is developed in the direction of the fold axes. Moine, Strath Farrar, Inverness-shire, Scotland

56B. Small folds of very regular amplitude and axial orientation have cylindrical hinges in the direction of a regional lineation (parallel to hammer handle). From the same locality as Plate 56A

Estwing

57A. A coarse crenulation lineation plunges vertically in mica schist. Paleozoic, near Randsverk, Oppland, Norway

57B. Mica schist has an almost perfectly linear fabric. Foliation is not obvious as a planar structure and the rock splits into rod-shaped fragments elongated in the direction of the regional lineation. Paleozoic, near Röros, Sör-Tröndelag, Norway

58. In massive schistose quartzite, lineation defined by faint striations plunging steeply right dominates a poorly defined foliation. Pre-Cambrian, near Alta, Finnmark, Norway

59A. A regular "mineral" lineation, defined by streaking and preferred orientation of elongated minerals, is present in quartz-schist. Pre-Cambrian, Papoose Flat, Inyo Mountains, California, U.S.A.

59B. In siliceous marbles a very prominent lineation is defined by streaks of calc-silicate minerals. Near Weldon, Sierra Nevada, California, U.S.A.

60. In massive gneissic schists a very prominent lineation plunges uniformly over a wide region. The lineation is defined largely by elongated streaks and aggregates of minerals. Petit St. Bernard, French Alps

61 A. In massive quartzo-feldspathoid gneiss a coarse lineation has a regular orientation over a wide region. It is defined by a ribbing and crenulation of a well developed foliation and by elongated aggregates of minerals. Turoka, Southern Kenya

61 B. From the same locality as Plate 61 A. Lineation is parallel to hammer handle

62 A. A very regular lineation is defined by the ribbing and grooving of foliation surfaces in flaggy quartz schist. Pre-Cambrian, near Alta, Finnmark, Norway

62 B. A quartz-schist has a strongly linear fabric defined by ribbing and grooving of the foliation surfaces. Near Mt. Pelion, Greece

63. Small folds, mullions and lineations plunge to the left in the hinge region of a large fold in quartz-mica schist. Moine, Glenshiel, Ross and Cromarty, Scotland

64A. In the hinge of a large fold in quartzo-feldspathic schist very regular lineation and incipient mullion structures plunge parallel to the fold axis. Near Vågå, Oppland, Norway

64B. The lineated hinge of a large fold plunges to the left. Moine, Kinlochmoidart, Inverness-shire, Scotland

65 A. Strongly developed mullion structure plunges parallel to the hammer handle in massive schistose quartzites. The rock breaks into stout log-shaped fragments with lineated surfaces. Moine, Strath Oykell, Sutherland, Scotland

65 B. Strongly developed mullion structure in massive impure quartzite. From the same locality as Plate 65 A

66 A. Large scale linear fabric developed in rocks with pronounced mullion structure. From the same locality as Plates 65 A and 65 B

66 B. Linear fabric in mylonitic quartzite with strongly developed lineation and mullion structure. In this rock many of the quartz grains are threadlike, elongated in the direction of the lineation. Near Barstow, California, U.S.A. (width of exposure about 2 m)

67 A. Weak open fold in alternating limestones and shales. Ordovician, Björkö, Oslo Fjord, Norway

67 B. Open concentric fold in bedded quartzites has one steep and one horizontal limb. Pre-Cambrian, Inyo Mountains, California, U.S.A.

68. Bedded limestones retain the same thickness in concentric monoclinal fold hinge. Jurassic, near Délémont, Jura, Switzerland

69. In a large fold in bedded quartzites some thicker layers are approximately concentrically folded. Others have sharp hinges at both surfaces and are more nearly similar in form. Dalradian, near Whitehills, Banffshire, Scotland (height of exposure about 7 m)

70. Layers of massive bedded quartzite show marked change in orthogonal thickness in hinge of large tight fold. The greatest thickness is in the axial surface, the least on the limbs. From the same locality as Plate 69

71 A. Anticline in massive bedded quartzite. Ordovician, Shap Fell, Westmorland, England

71 B. From the same locality as Plate 71 A

72 A. Folds in interbedded sandstones and shales have sharp hinges and planar limbs. Carboniferous, near Sandymouth, Devonshire, England

72 B. Advanced massive erosion in the hinge region of a fold in interbedded sandstones and shales. Carboniferous, near Hartland, Devonshire, England

73. Bedded sandstones contain small folds with sharp hinges and planar limbs. Permo-Carboniferous, near Matesevo, Montenegro, Yugoslavia

74A. Recumbent fold in flysch. Cretaceous, Lidorikion, Greece

74B. Thicker sandstone layers are folded concentrically. From the same locality as Plate 74A

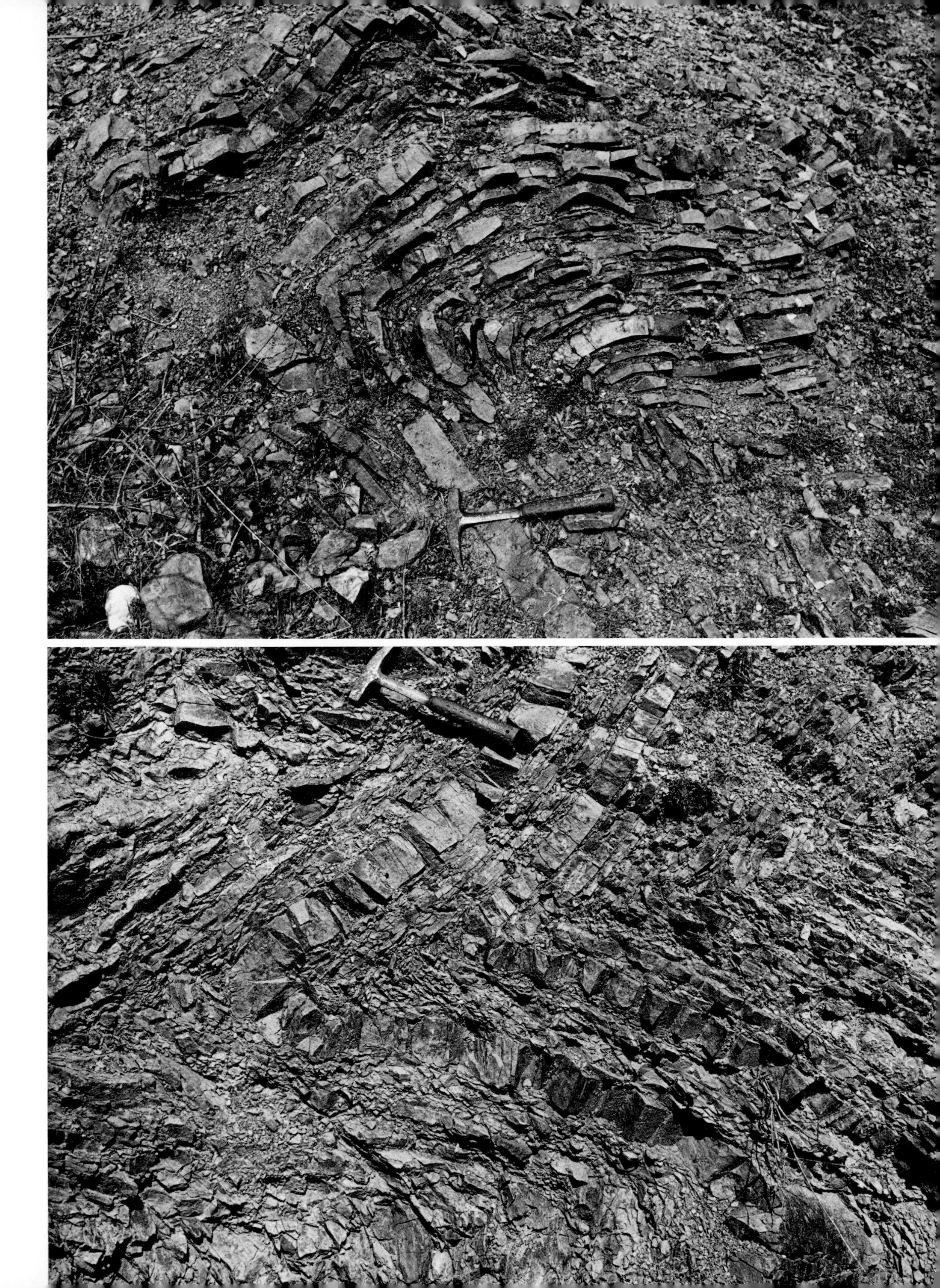

75. The hinge of a large fold in a sandstone layer plunges toward the camera. The folded surface forms an almost circular arc in profile, and the limbs are planar. Carboniferous, near Hartland, Devonshire, Endland

76A. Interbedded sandstones and shales contain large "chevron" folds with sharp hinges and planar limbs. Axial planes of the folds dip very gently away from the observer. Carboniferous, Millook Haven, Devonshire, England (scale is given by figures, lower left)

76B. Tight folds in interbedded sandstones and shales have steeply dipping axial surfaces. Carboniferous, north of Bude, Devonshire, England (height of exposure about 25 m)

77A. Detail of Plate 76A. In the fold closing to the left (center) beds on the upper limb are thinned and crossed by quartz-filled extension fractures

77B. Detail of Plate 77A (lower right). In this fold beds in the lower limb are preferentially thinned

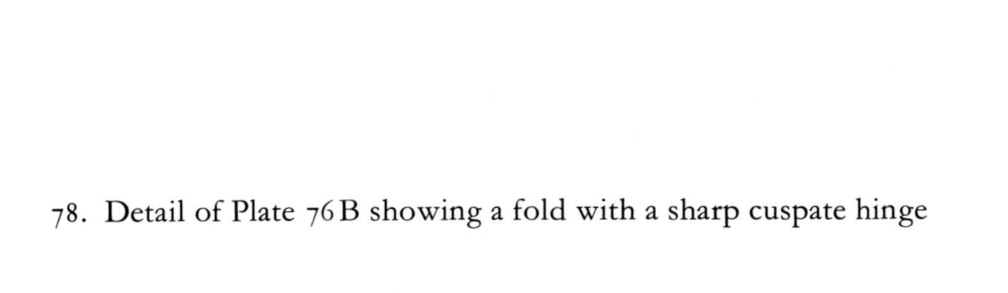

78. Detail of Plate 76 B showing a fold with a sharp cuspate hinge

79. A "box" or conjugate synformal fold is present in massive bedded quartzite. Pre-Cambrian, Inyo Mountains, California, U.S.A.

80 A. Cylindrical folds plunging gently to the left are present in massive graywackes and siltstones. A pronounced cleavage is developed parallel to their subvertical axial planes. Pre-Cambrian, South Stack, Holy Island, Wales

80 B. Folded interbedded shales and cherts. Franciscan, Marin County, California, U.S.A.

81. Folds in shale layers (darker color) and chert layers have different forms in profile. From the same locality as Plate 80B

82 A. Massive bedded limestones contain a very large fold with sharp hinge and planar limbs. Cretaceous, Lake Lucerne, Switzerland (scale given by buildings at lower edge)

82 B. From the same locality as Plate 82 A (scale given by buildings upper left)

83. Gently plunging folds in bedded graywackes have distorted profiles in horizontal exposure surface. Paleozoic, near Bateman's Bay, New South Wales, Australia

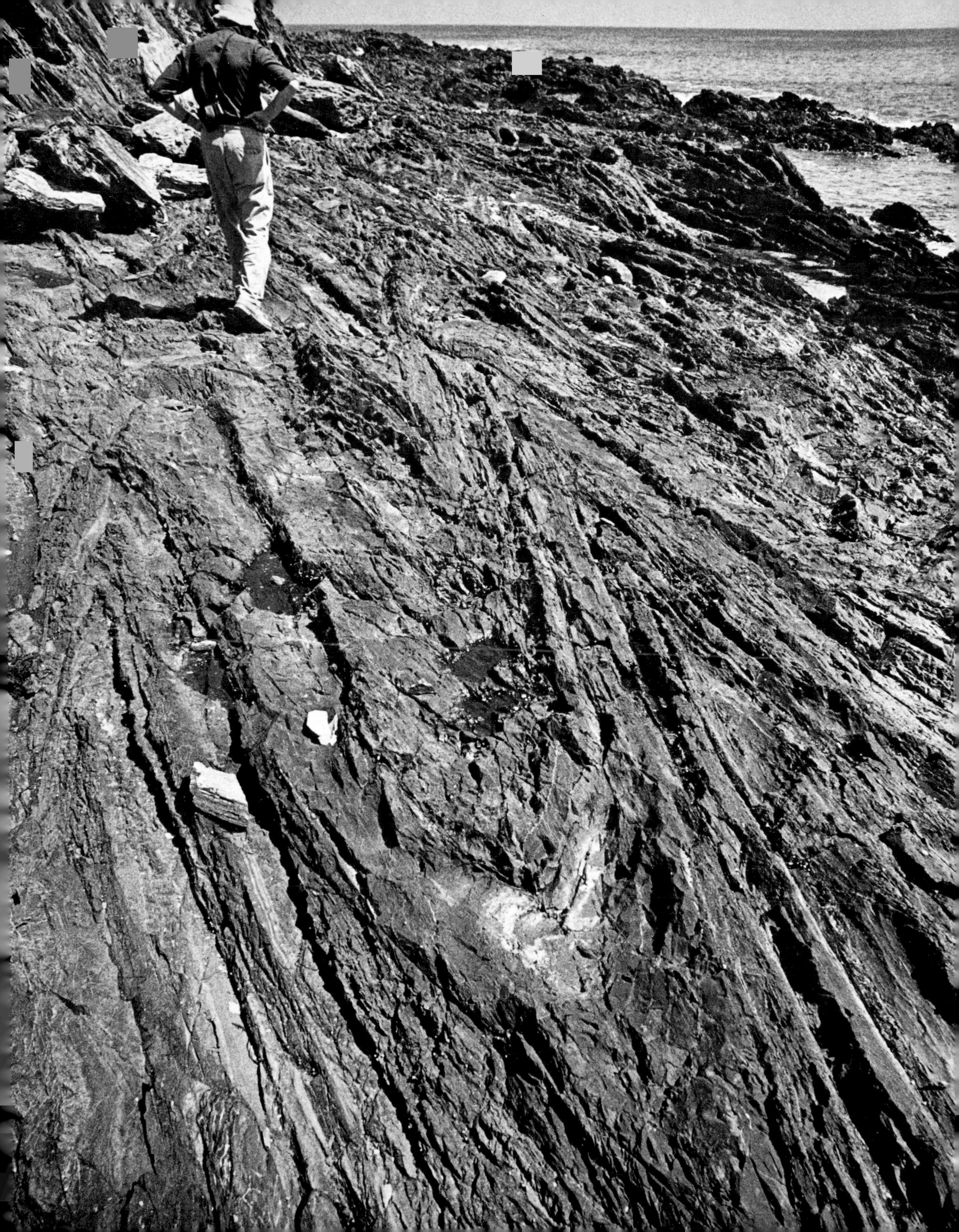

84A. Disharmonic folding of graywacke beds involves marked thinning of some fold limbs. From the same locality as Plate 83 (height of exposure about 2 m)

84B. Tight folds in graywackes have sparse cleavage fractures subparallel to axial planes. From the same locality as Plates 83 and 84B

85 A. A rounded fold in a massive quartzite bed has fan-like cleavage fractures inclined symmetrically to its axial plane. The lower surface of the layer is cuspate in form and the underlying mudstone penetrates the quartzite along a large fracture at the hinge. Dalradian, near Whitehills, Banffshire, Scotland (see also Plate 1 B)

85 B. Beds of graywacke are greatly thickened and traversed by cleavage fractures in the hinge of a fold. Silurian, near Kirkudbright, Scotland

86 A. A tightly-folded massive quartzite is traversed by a fan of cleavage fractures. A more pervasive foliation is developed in the surrounding phyllitic shales. Pre-Cambrian, South Stack, Holy Island, Wales

86 B. Local erosion has removed phyllitic layers to expose vertically plunging cylindrical folds in thin layers of quartzite. Dalradian, near Whitehills, Banffshire, Scotland

87. Large folds in bedded graywackes and siltstones have subvertical axial plane cleavage. Pre-Cambrian, South Stack, Holy Island, Wales (height of exposure about 40 m)

88. Thin bedded sandstones contain regular kink-like folds. In thicker layers, hinges are concentric in form. Permo-Carboniferous, near Matesevo, Montenegro, Yugoslavia

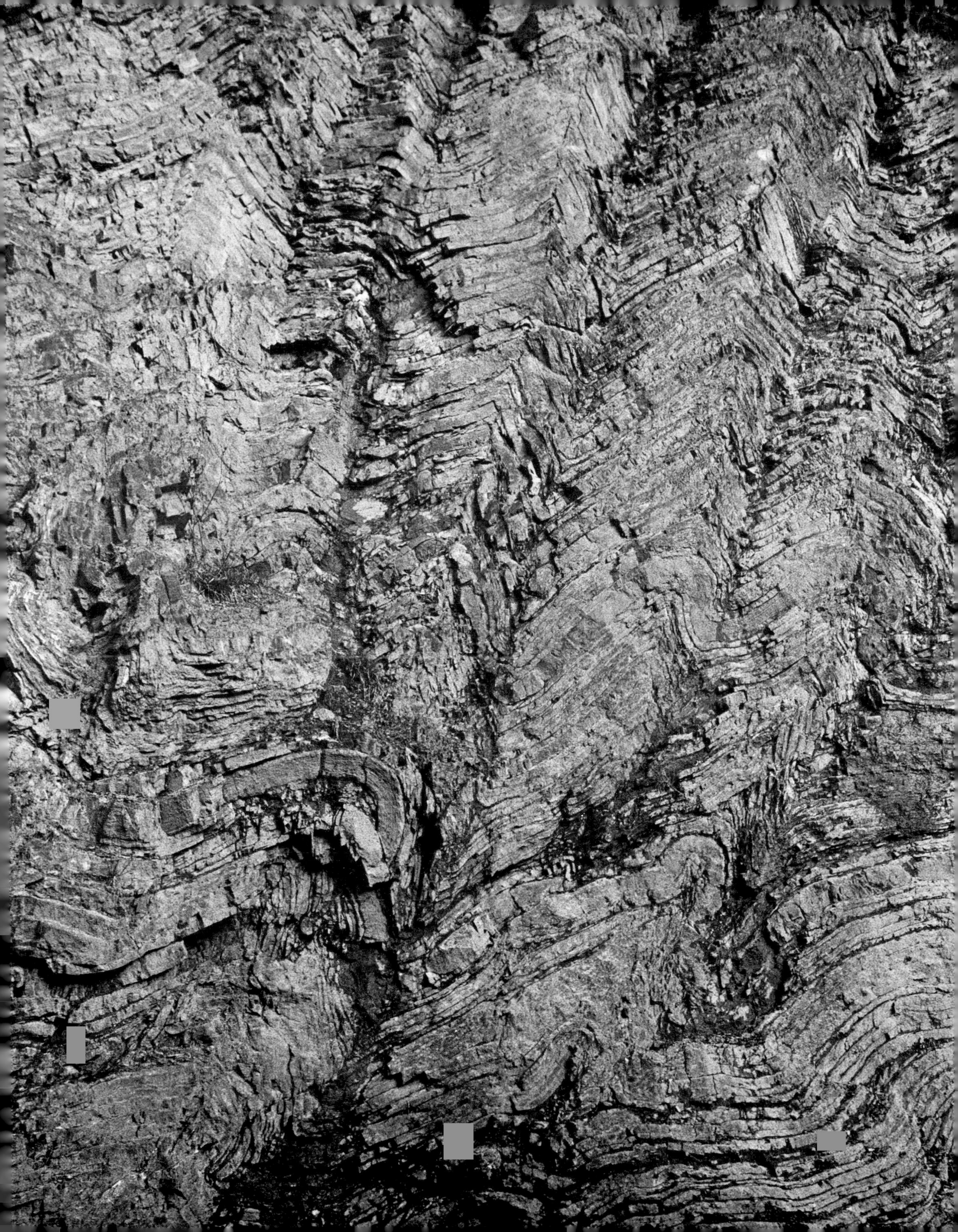

89. Tight concentric folds are present in an isolated quartzite bed enclosed in crenulated mica schist. Dalradian. Lock Leven, Inverness-shire, Scotland (height of exposure about 5 m)

90A. Thin beds of quartzite are isoclinally folded in intensely crenulated mica-schist. Two orders of folding are clearly visible in the quartzite layers. From the same locality as Plate 89

90B. Large folds in quartzite layers in mica schist. From the same locality as Plates 89 and 90A

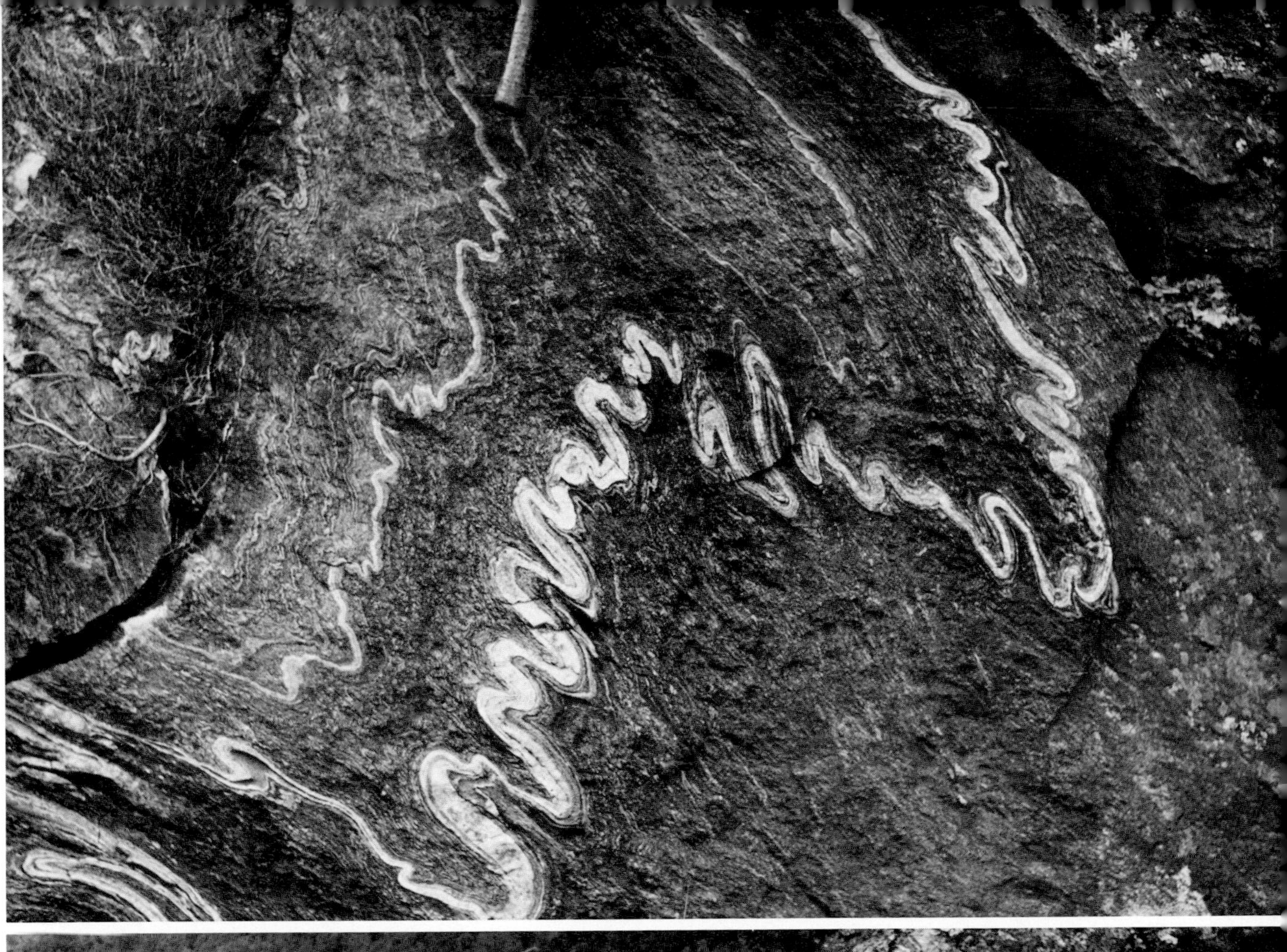

91 A. Tight folding of layering in impure marble. Tsangarada, near Volos, Greece

91 B. Thin folded quartzite layers in mica schist. Near Oppdal, Oppland, Norway

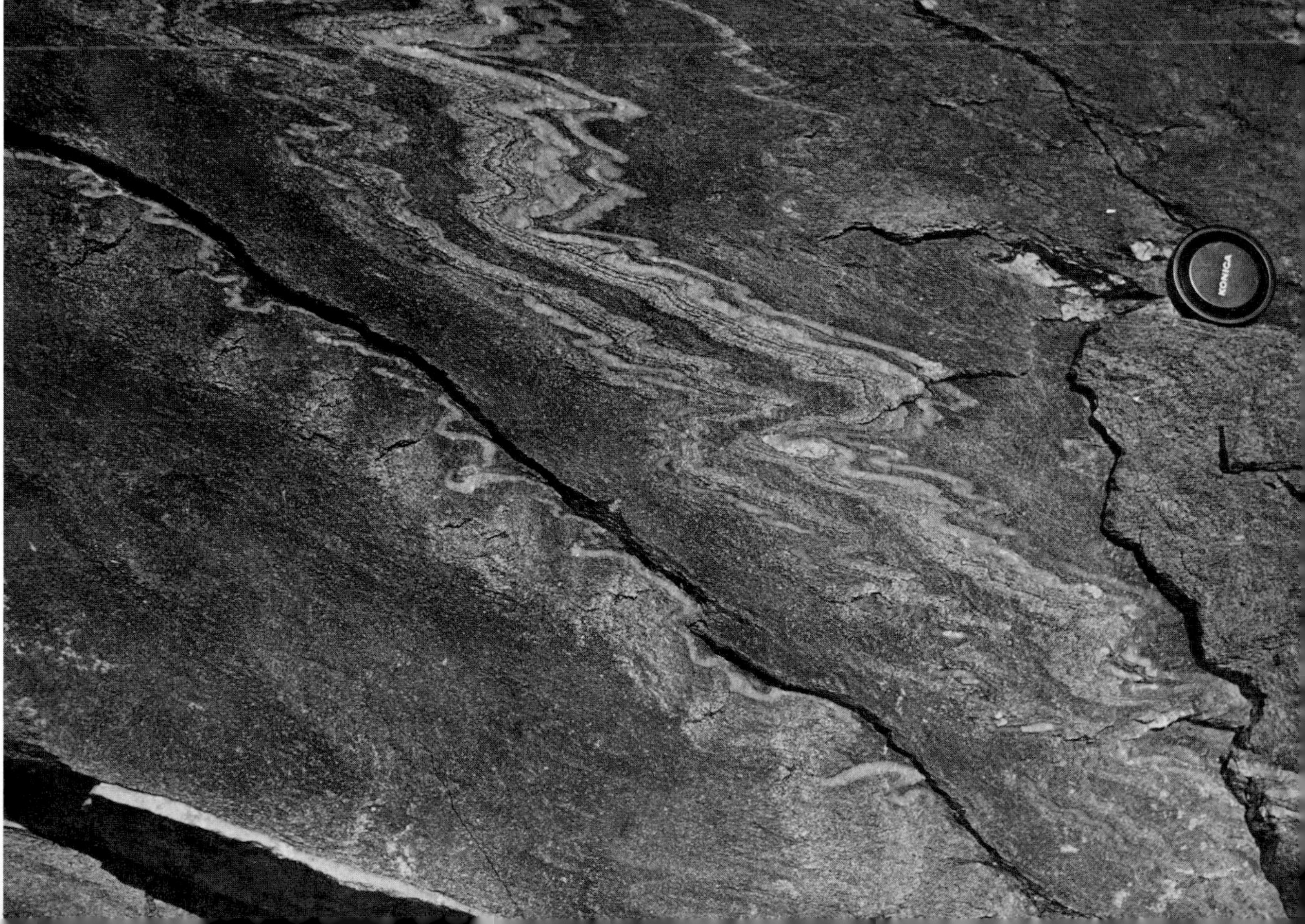
KONICA

92 A. Folded layers in quartzo-feldspathic gneiss are markedly thickened in the fold hinges. Near Machakos, Kenya

92 B. Isoclinal folding in gneiss. From the same locality as Plate 92 A

93 A. One limb of a fold in a quartzo-feldspathic layer in gneiss is grossly attenuated. Lewisian, South Uist, Outer Hebrides, Scotland

93 B. Folded surfaces are approximately sinusoidal in profile. Increase in orthogonal thickness of folded layers occurs in the hinge. Moine, Strath Farrar, Inverness-shire, Scotland (width of exposure about 3 m)

94 A. Folding is tight and approximately similar in profile. An exposed hinge line (center) gives the direction of plunge. Moine, Kinlochmoidart, Inverness-shire, Scotland

94 B. A similar fold is present in massive quartzo-feldspathic schists. From the same locality as Plate 94 A

95 A. Tight similar folding in laminated mylonite as seen in profile. Near Musgrave Park, Musgrave Range, South Australia

95 B. Tight recumbent folds in quartzite interlayered with thin mica schists show extreme increase in orthogonal thickness of layers in the hinges. Direction of plunge is given by an exposed hinge line (center). Ord-Ban, Strathspey, Scotland

96. Approximately similar fold in gneiss. Pre-Cambrian, south of Namsos, Norway

97 A. Fold in massive quartzo-feldspathic schist showing markedly different curvature from layer to layer. Moine, Glenfinnan, Inverness-shire, Scotland

97 B. Fold in bedded quartzites. Lower Paleozoic, near Narvik, Nordland, Norway

98 A. Fold in bedded graywackes with tightly appressed (cuspate) core. This fold is more concentric than similar. Lower Paleozoic, near Randsverk, Oppland, Norway

98 B. Cylindrical reclined folds in quartz-mica schists. The fold axes are parallel to prominent lineation and mullion structure in surrounding schists. Moine, Strath Oykell, Sutherland, Scotland

99 A. Small folds in metamorphosed thin-bedded cherts show local attenuation of layers on limbs. Near Sutter Creek, California, U.S.A.

99 B. Detail of Plate 99 A (upper center)

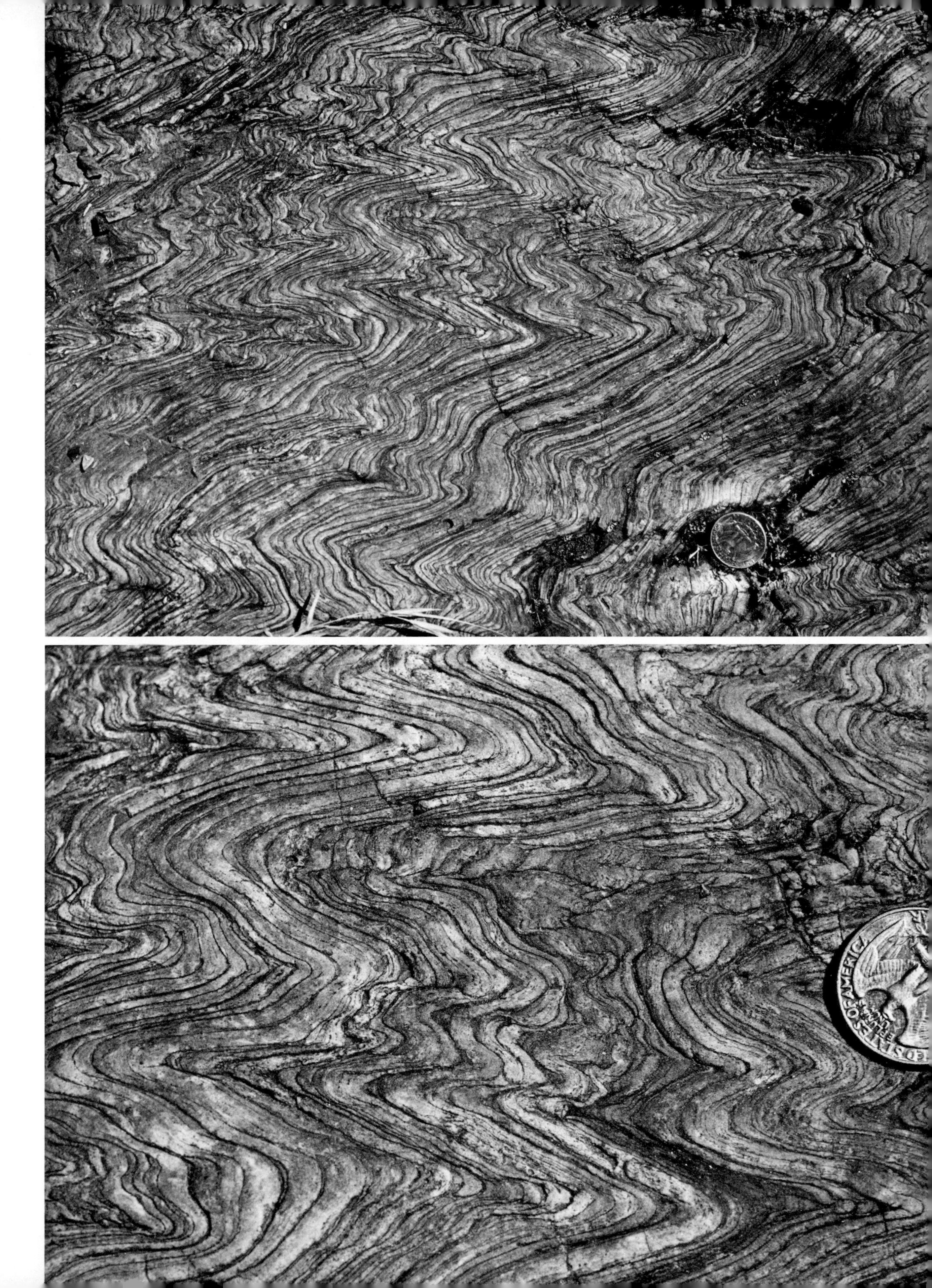

100 A. Extreme attenuation of limbs is present in some folds. Locally, limbs are replaced by incipient transposition surfaces subparallel to the axial planes of the folds. From the same locality as Plates 99 A and 99 B

100 B. Tight sharp-hinged folds in platy quartz schist. Pre-Cambrian, Oppdal, Oppland, Norway

101. Folds in Sparagmite show features similar to those of folds in Plates 99 A, 99 B and 100 A. Near Gjendesheim, Oppland, Norway

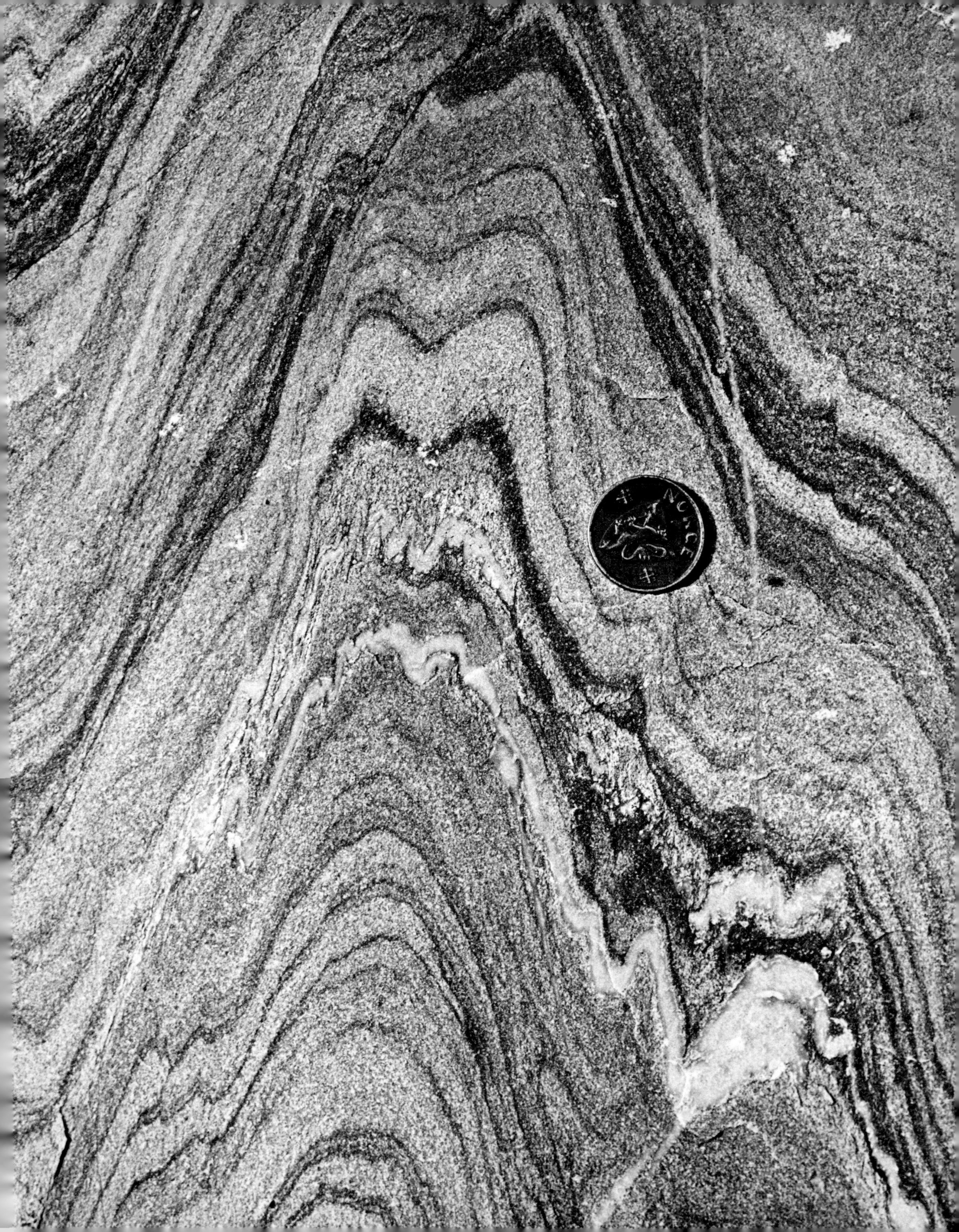
NORGE

102 A. Banded marbles contain folds of markedly similar style. Limbs are planar and hinges are sharp. Dalradian, near Tomintoul, Banffshire, Scotland

102 B. Markedly cylindrical similar folds in impure marble. From the same locality as Plate 102 A

103 A. Tight folds in impure marble. From the same locality as Plates 102 A and 102 B

103 B. Folded Moine schists, Kyle of Tongue, Sutherland, Scotland

104. Small similar folds in quartzo-feldspathic schist. Moine, Strath Farrar, Inverness-shire, Scotland

Ever Grip

105 A. Folded quartzo-feldspathic veins are present in mica schist with uniform foliation parallel to axial planes of the folds. Near Bygdin, Oppland, Norway

105 B. Relict fold hinges survive locally in a foliated *mélange*. Lower Paleozoic, near Narvik, Nordland, Norway

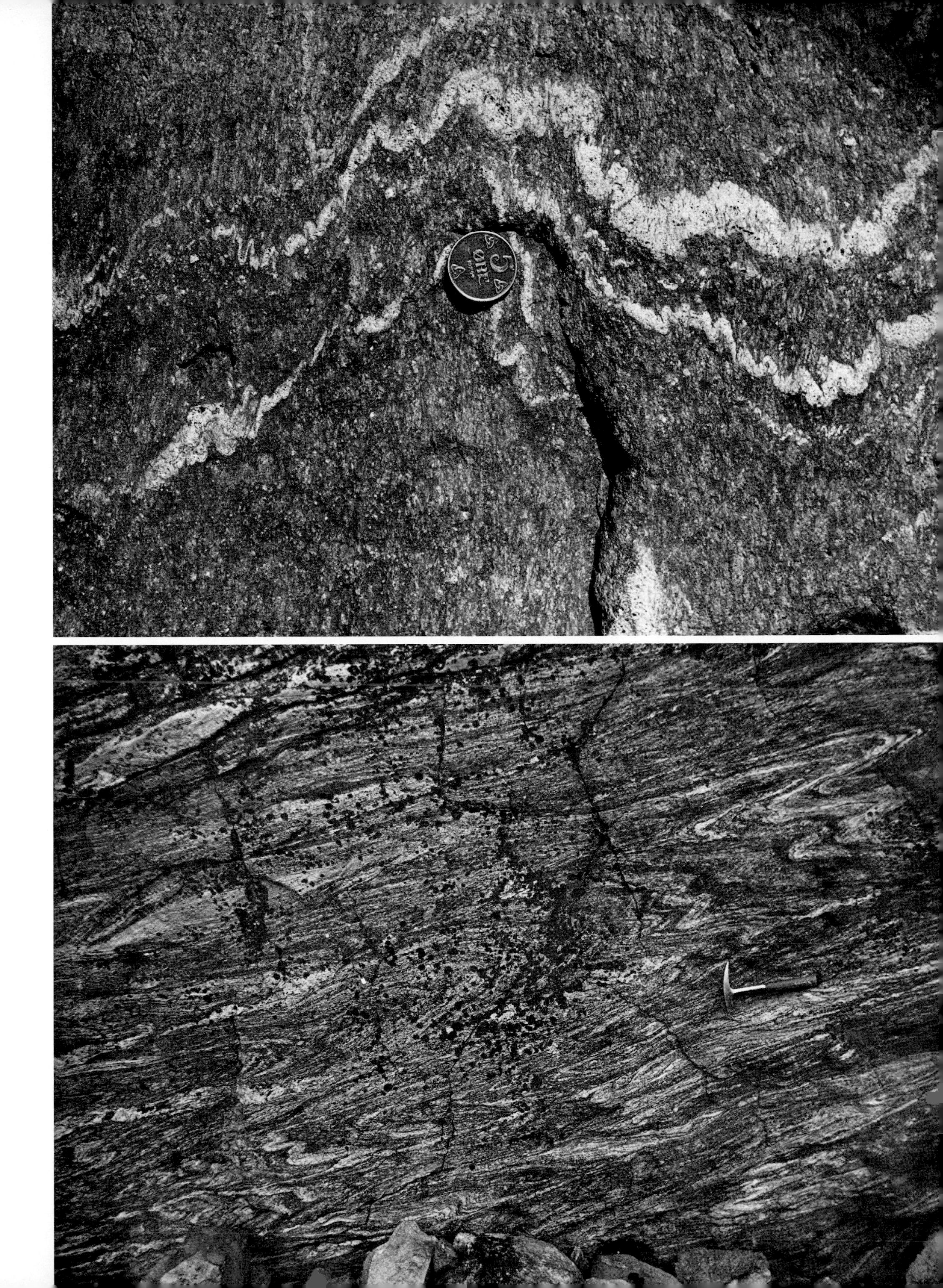
5 ØRE

106 A. Isoclinal concentric fold in quartzo-feldspathic layer in biotite gneiss. Pre-Cambrian, near Namsos, Nord-Tröndelag, Norway

106 B. Quartzo-feldspathic veins in biotite gneiss contain folds of various sizes. Moine, Glen Cannich, Inverness-shire, Scotland

5
ØRE

107. Relict isoclinal fold in quartzo-feldspathic schist is shown distorted in profile obliquely inclined to the fold axis. Moine, near Lochailart, Inverness-shire, Scotland

108. From the same locality as Plate 107

109. Relict folds in quartzo-feldspathic layers in biotite gneiss. Pre-Cambrian, near Fagernes, Oppland, Norway

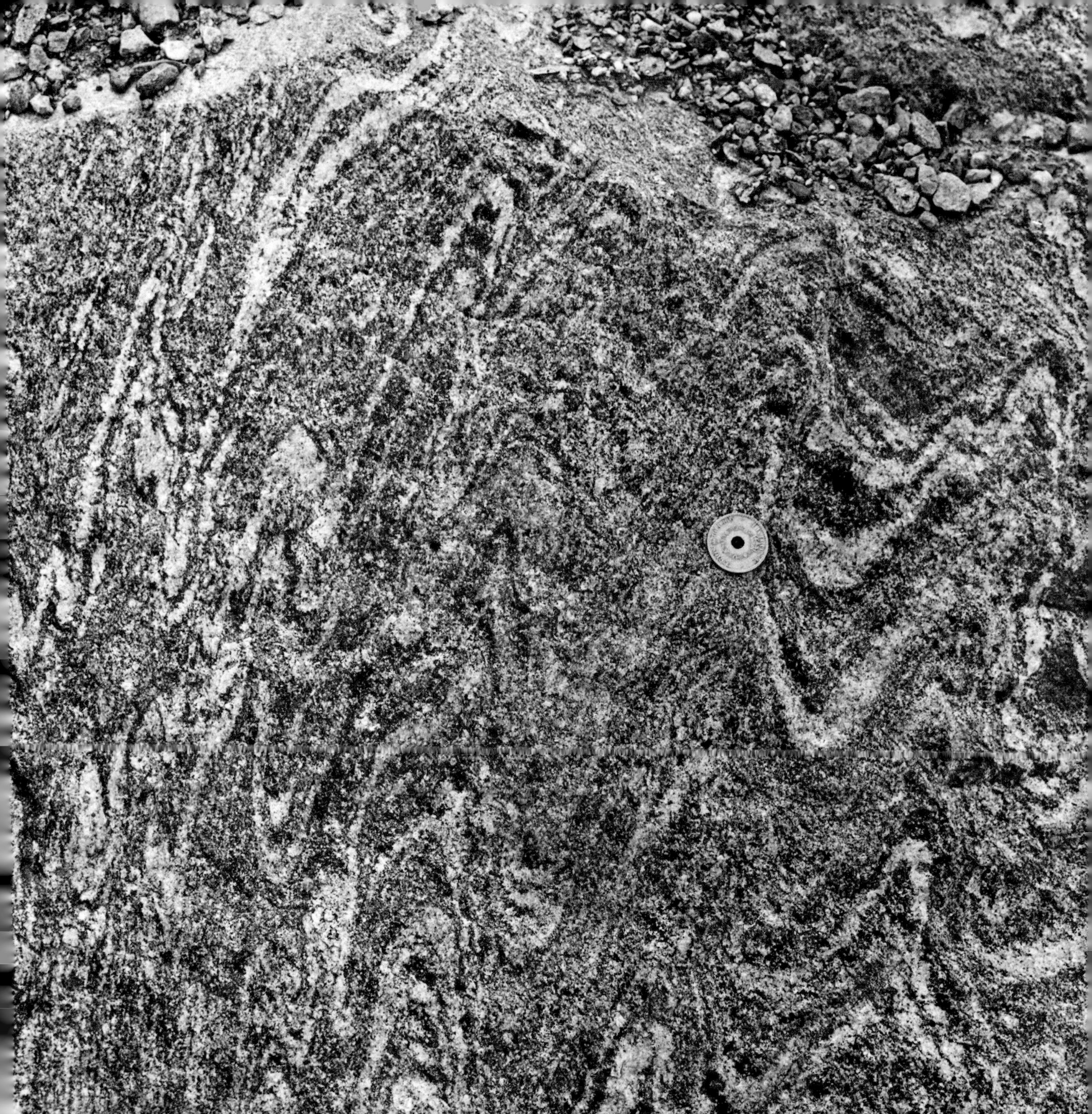

110 A. Relict fold in schist *mélange*. South of Kavala, Macedonia, Greece

110 B. Fold in impure marble. Tsangarada near Volos, Greece

111 A. Angular similar-style folds are preserved in faint colored laminae in marble. Near Weldon, Sierra Nevada, California, U.S.A.

111 B. The origin of these similar fold-like structure in quartzite is unclear. They could be of tectonic origin, or they could be deformed sedimentary structures. Paleozoic, near Cooma, New South Wales, Australia

112A. Open angular kink-folds are present in the foliation of a phyllite. An earlier lineation is perpendicular to the axis of the kink-folds. Moine, near Achnasheen, Ross and Cromarty, Scotland

112B. Open angular kink-folds in phyllite. Col du Lauteret, French Alps

Estwing

113. Kink folds of opposite senses are present in the foliation of a phyllite. The hinges of the two sets are not parallel and the conjugate kink folds formed by each pair are noncylindrical. From the same locality as Plate 112A

114. A kink-fold (or kink "band") displaces the foliation of a phyllite.
Devonian, Ilfracombe, Devonshire, England

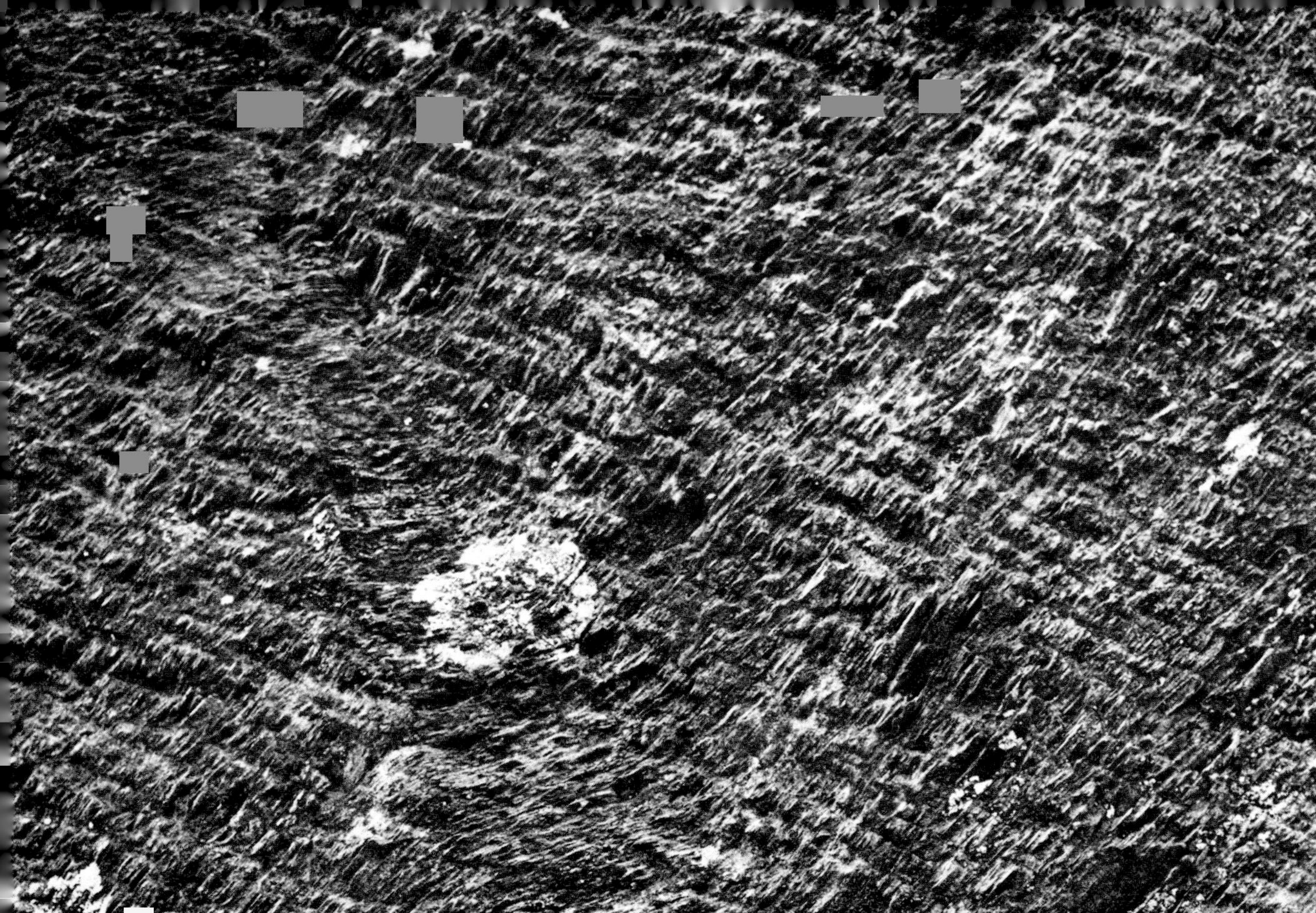

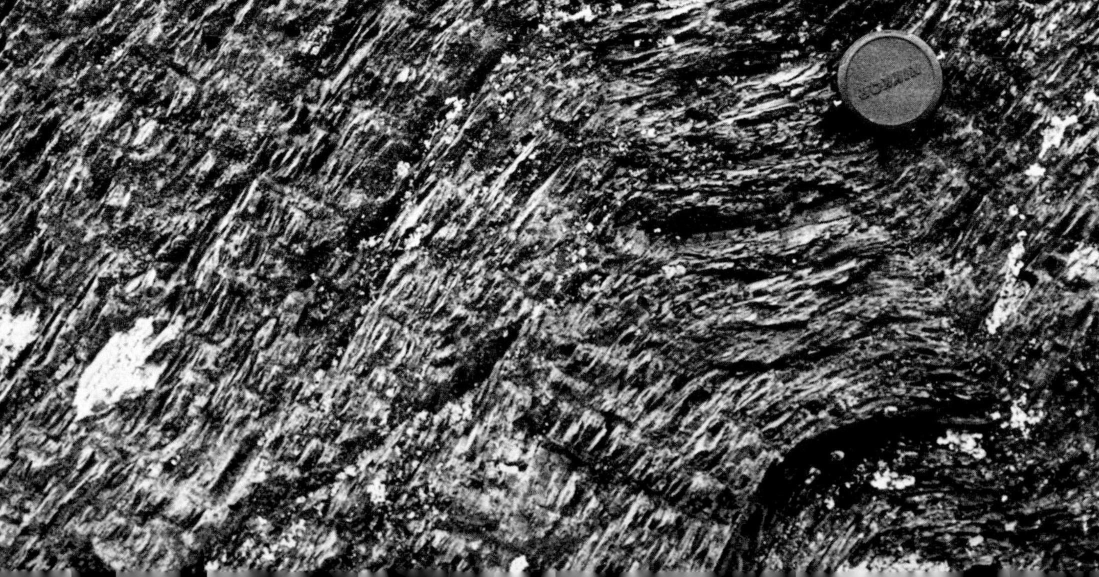

115 A. Open kink-folds in phyllite. Carboniferous, near Brest, Brittany, France

115 B. A small conjugate kink-fold affects transposition foliation in phyllite. Relict folded bedding can be seen in some of the thicker foliation laminae (upper left). From the same locality as Plate 115 A

Estwing

116. Quartz-filled fractures are present in a kink band in phyllite. Fractures occur along the boundaries of the band and between foliation surfaces within the band. From the same locality as Plates 115 A and 115 B

117A. Narrow kink bands in phyllite are parallel sided and regularly spaced. Some are bounded by planar fractures. Devonian, Morthoe, Devonshire, England

117B. Narrow kink bands have curved boundaries. Some are lenticular. From the same locality as Plate 117A

EverGrip

118. Narrow kink bands have strongly curved boundaries marked by open fractures. From the same locality as Plates 117A and 117B

119A. Kink bands with rounded hinges in phyllite. Pre-Cambrian, Holyhead, Holy Island, Wales

119B. Kink bands with rounded hinges in quartz-phyllite. Tre-Arddur Bay, Holy Island, Wales

120. Kink band in biotite gneiss is narrow with ill-defined boundaries and rounded hinges. Near Cannobio, Lake Maggiore, Italy

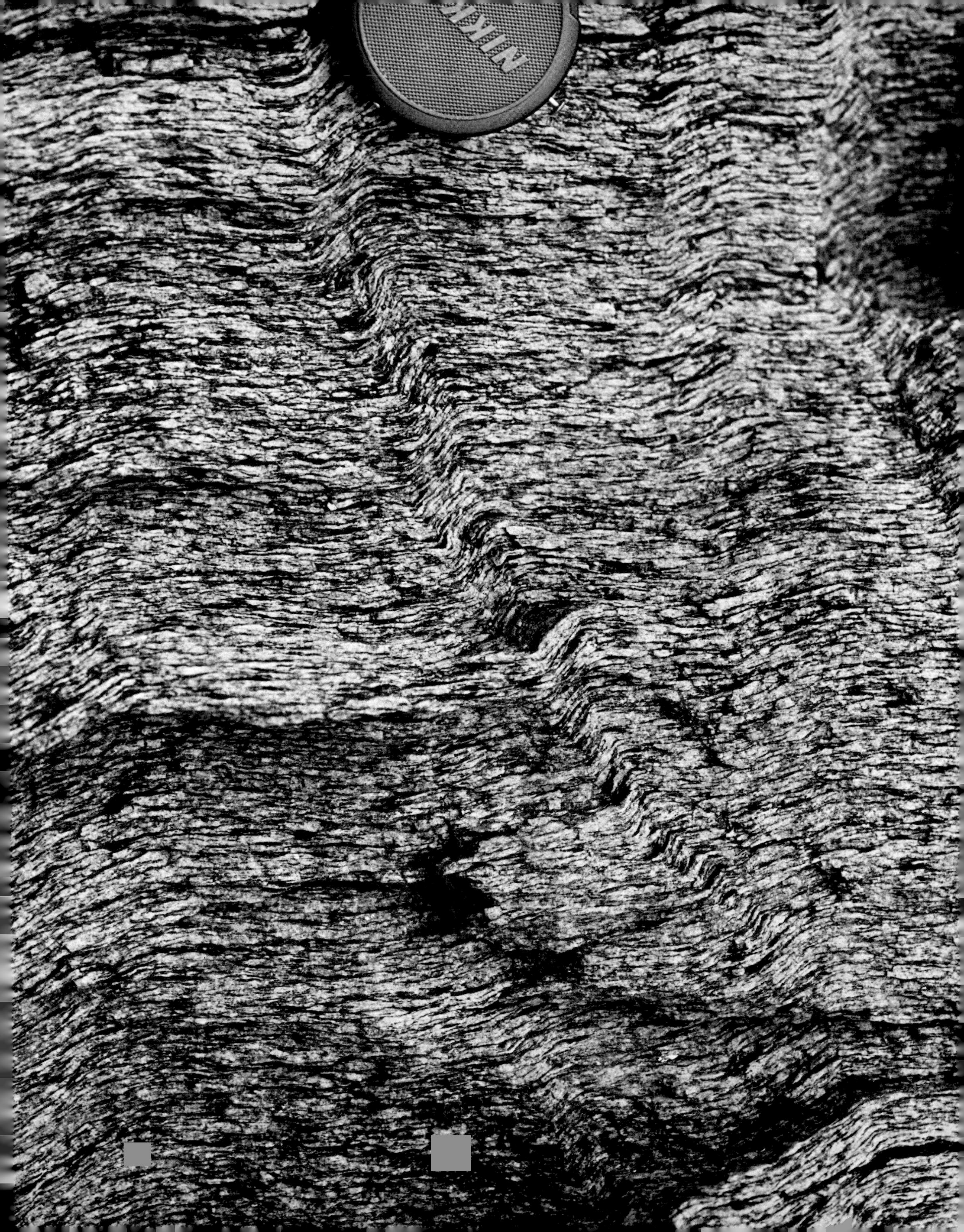

121. Conjugate kink bands with rounded hinges form an intersecting network in biotite gneiss. Intersections of two bands with associated conjugate folding can be seen in several places. From the same locality as Plate 120

122. Asymmetric conjugate kink bands form an intersecting network in phyllite. Pre-Cambrian, Tre-Arddur Bay, Holy Island, Wales

123 A. Large conjugate fold in phyllite (left). Lower Paleozoic, Stjördal, Nord-Tröndelag, Norway

123 B. Incipient folds in interbedded shales and cherts take the form of weak kink-like bands dipping left. Tertiary, Berkeley Hills, California, U.S.A. (height of exposure about 5 m)

124A. A large round-hinged kink fold dips gently right in flaggy laminated quartz schists. The foliation curves markedly outside the band. Pre-Cambrian, Oppdal, Oppland, Norway

124B. A small incipient kink band dips steeply to the right. From the same locality as Plate 124A

125 A. Parallel sets of large kink bands (dipping right) are associated with incipient refolding of flaggy quartz schist. From the same locality as Plates 124 A and 124 B (height of exposure about 25 m)

125 B. A large kink band developed during incipient folding of massive bedded limestones. Cretaceous, Morača River, Yugoslavia

126. Smaller bands of kink-like chevron folds are developed in the hinge and on the limbs of a large fold in phyllite. From the same locality as Plate 123 A

127. Small-scale chevron folding of a phyllitic foliation takes the form of horizontal parallel sided kink bands. From the same locality as Plates 123 A and 126

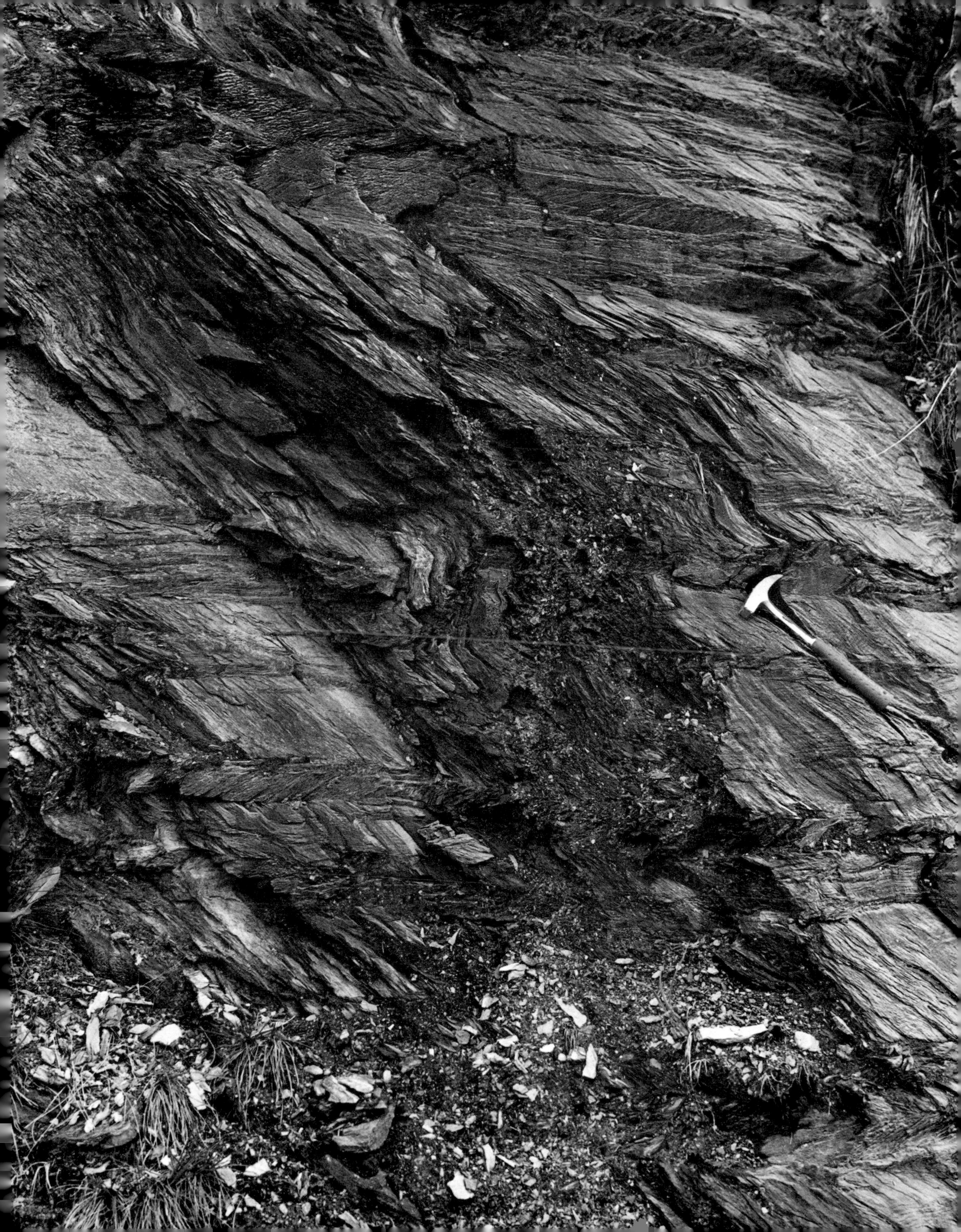

128. Fold in schist with planar limbs and rounded hinge. Lower Paleozoic, near Storlien, Sweden

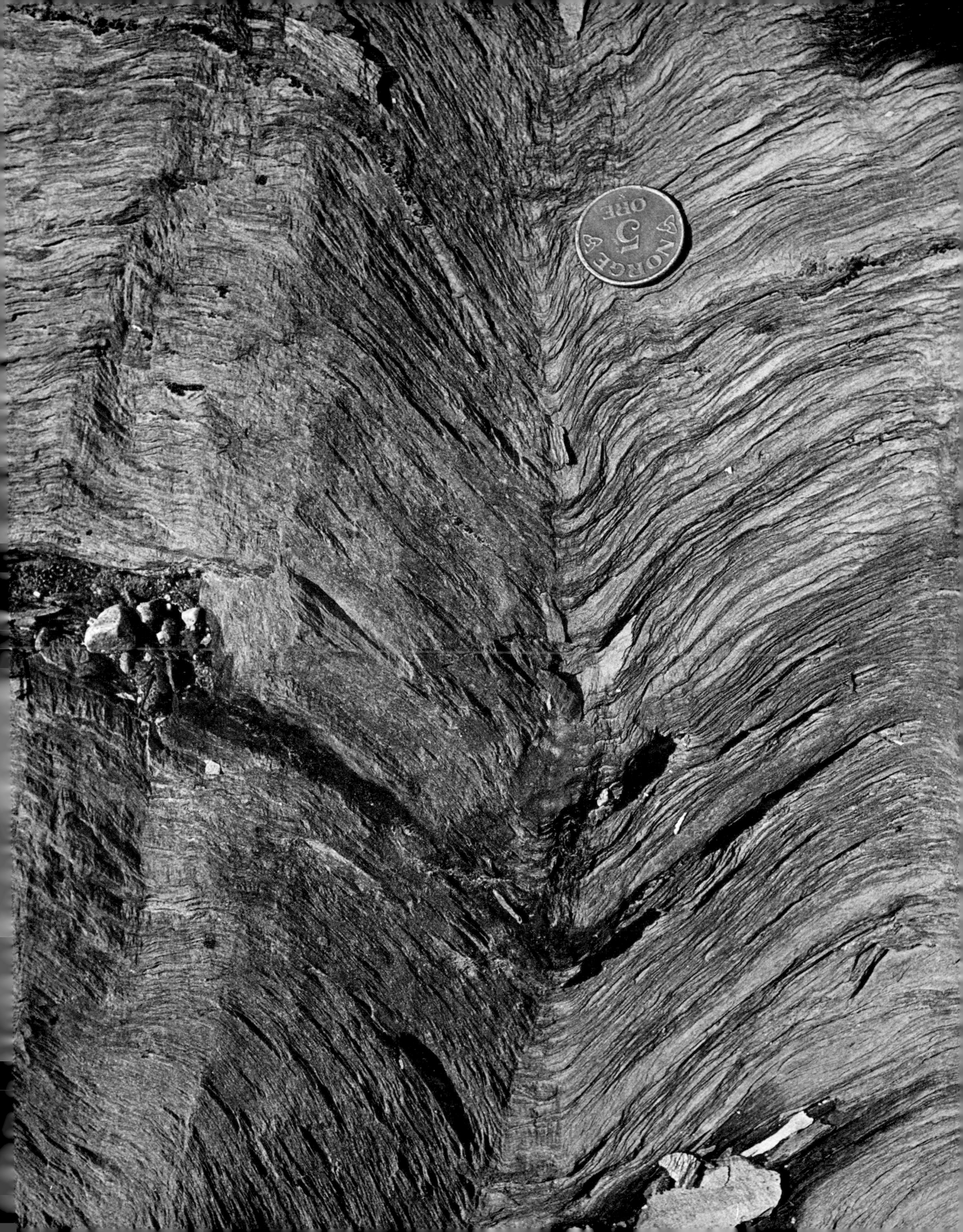
5
ØRE
NORGE

129. Chevron folding in schist. From the same locality as Plate 128

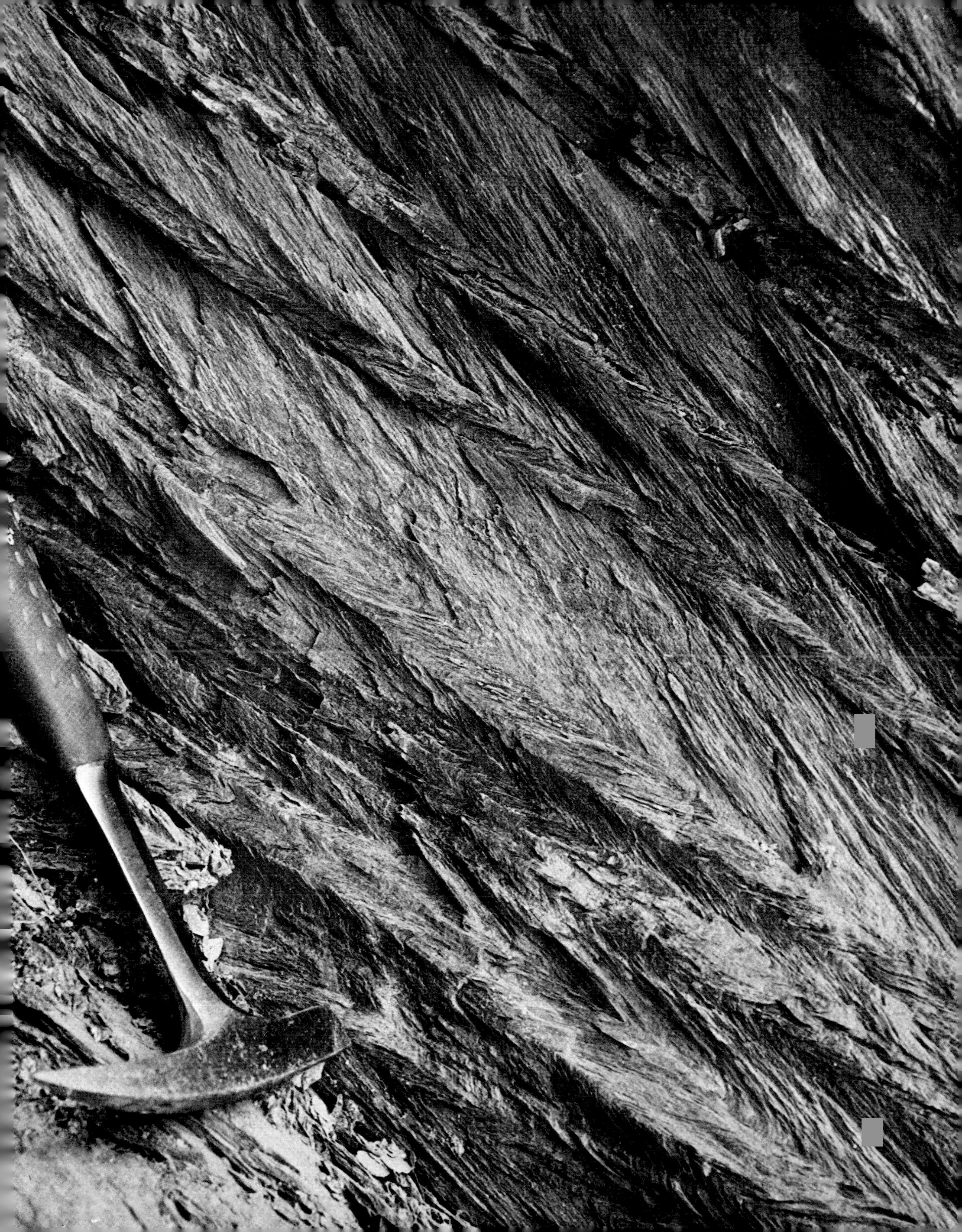

130A. Small folds in phyllite range from kink bands with angular hinges (upper left) to rounded forms. Pre-Cambrian, Tre-Arddur Bay, Holy Island, Wales

130B. From the same locality as Plate 130A

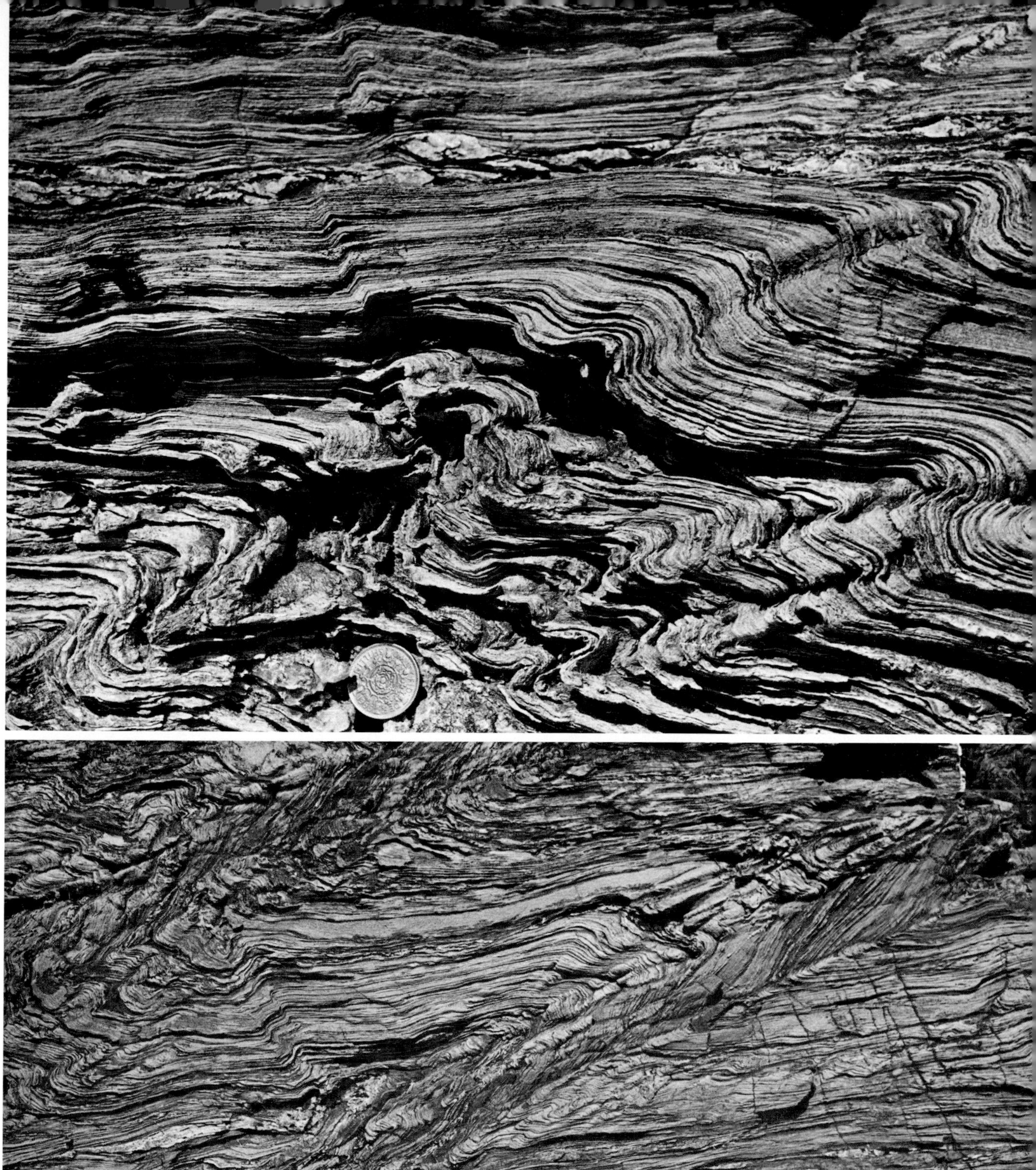

131. Folding in interlayered quartzite and phyllite. From the same locality as Plates 130A and 130B

132. Folding in quartz-phyllites. Pre-Cambrian, Holyhead, Holy Island, Wales

133. Cylindrical crenulations in phyllite. Lower Paleozoic, Stjördal, Nord-Tröndelag, Norway

134. Cylindrical crenulations or chevron folds in phyllite are crossed by a weak earlier lineation (upper left). From the same locality as Plate 133

135. Hinges of chevron folds in phyllite plunge gently left. Dalradian, Lock Eck, Argyllshire, Scotland

136A. Weak open crenulations (plunging left) in the foliation surface of a schist are crossed by faint earlier lineation (plunging right). Moine, Loch Shiel, Inverness-shire-Scotland

136B. Strongly kinked or crenulated foliation in phyllite contains an early lineation perpendicular to the hinges of the kink folds. Devonian, Les Eaux-Bonnes, Pyrenées, France

137A. Chevron folds with sharp kink-like axial surfaces are present in quartz-schists. Pre-Cambrian, near Rhoscolyn, Holy Island, Wales

137B. Folding in phyllite. Dalradian, near Dunoon, Argyllshire, Scotland

138. In phyllite folds are defined by bands of kinking. Locally (center, above hammer head) conjugate kink bands can be seen. Dalradian, Lock Eck, Argyllshire, Scotland

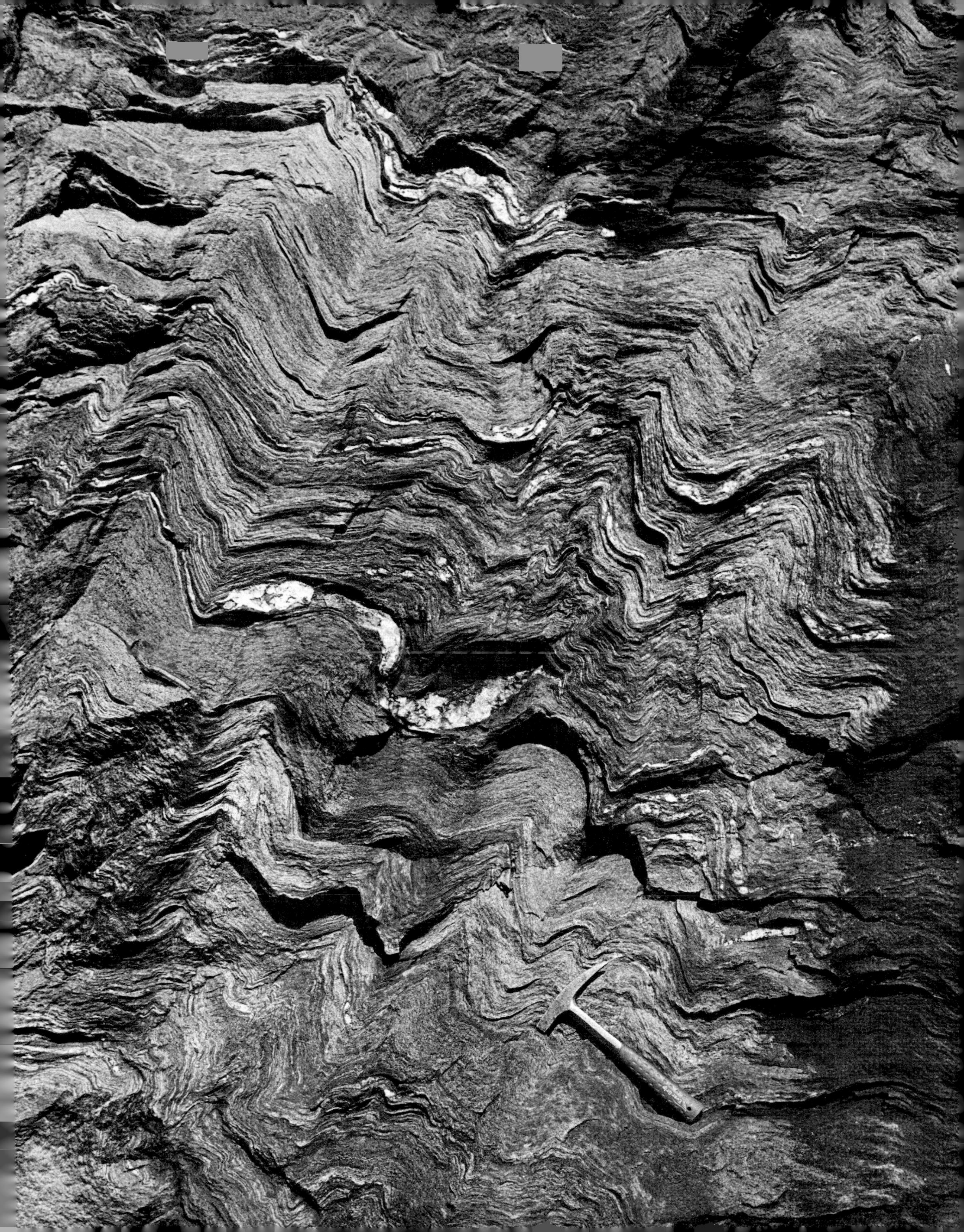

139. Chevron crenulations in phyllite are defined by very thin kink bands. Locally (upper right) conjugate sets can be seen. Paleozoic, Stjördal, Nord-Tröndelag, Norway

140 A. Conjugate fold in phyllite defined by intersecting kink bands. From the same locality as Plate 138 (height of exposure about 1 m)

140 B. Tight folding of siliceous layers in impure marble. Mount Pelion, Greece

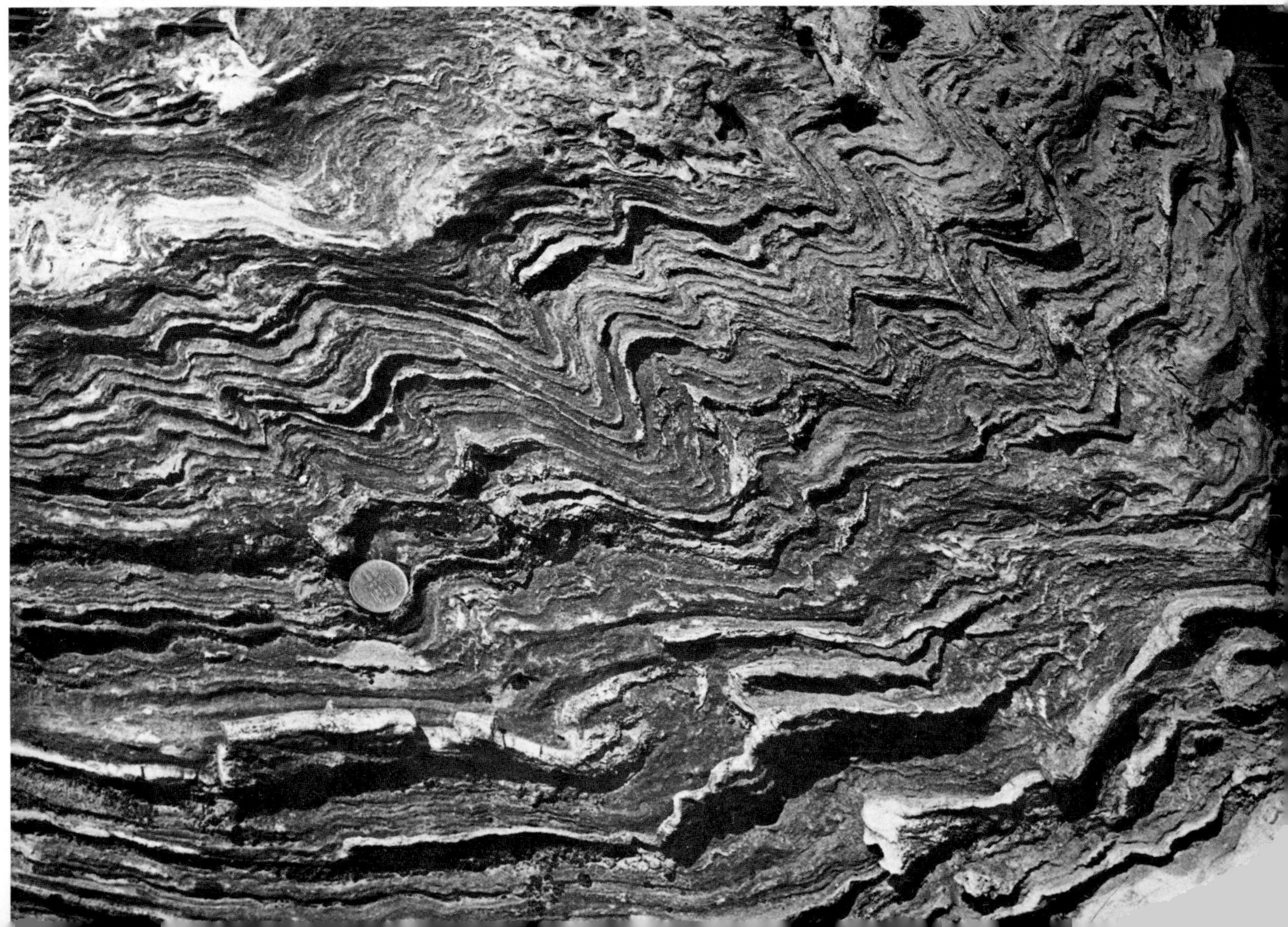

141 A. Fracturing and incipient rotation of angular fragments of quartzo-feldspathic layer in gneiss. South of Mosjöen, Nordland, Norway

141 B. Boudins of graywacke in shale contain quartz-filled extension fractures. Bacco Pass, near Rapallo, Italy

142 A. Dolomite rock layers in marble are boudinaged by necking and separation. Near Tromsö, Troms, Norway

142 B. From the same locality as Plate 142 A

NORGE
1 KRONE
19 57

143. Boudins of massive graywacke in shale are angular and appear to have been formed largely by fracture of a thick bed. Near Naufpaktos, Greece

15

144 A. Angular boudins of amphibolite in quartzo-feldspathic gneiss. Lewisian, South Uist, Outer Hebrides, Scotland

144 B. Calc-silicate layers in marble have been necked and separated into boudins. Near Weldon, Sierra Nevada, California, U.S.A.

145 A. A granitic layer in mica schist has been separated into ellipsoidal boudins. Near La Viella, Pyrenées, Andorra

145 B. Boudinage of granitic layer in schist. Paleozoic, near Fauske, Nordland, Norway

146 A. Fractured boudins of quartzite are aligned in the foliation of schist. Dalradian, near Portsoy, Banffshire, Scotland

146 B. Boudin of massive quartzite in strongly foliated schist and quartzite. Pre-Cambrian, Inyo Mountains, California, U.S.A.

Estwing

147A. In marble thin siliceous layers are disrupted and boudinaged during development of a prominent foliation. The large light-colored boudins are fragments of a thin granitic body. Near Moesjöen, Nordland, Norway

147B. Foliation in marble passes smoothly round boudins of granite. From the same locality as Plate 147A

148 A. Eye-shaped quartzo-feldspathic boudin in laminated gneiss. Near Alta, Finnmark, Norway

148 B. Large boudins of granite in biotite gneiss, near Machakos, Kenya

Estwing

149 A. A quartzo-feldspathic vein obliquely inclined to foliation in schist is separated into boudins by very local necking and "pinching off." Lower Paleozoic, near Korgen, Nordland, Norway

149 B. A cross-cutting vein in crenulated gneiss is separated into short ellipsoidal boudins, some of which appear to have rotated. Lewisian, Loch Duich, Ross and Cromarty, Scotland

150. Boudinage of quartzo-feldspathic layers conformable to foliation in basic gneiss. South of Moesjöen, Nordland, Norway

151. Thin vein in basic layer in gneiss is separated into boudins, whereas adjacent quartzo-feldspathic layers are not. From the same locality as Plate 150

152. A massive graywacke layer in shale is separated into large boudins with little internal deformation. The lateral displacement of the boudins is a result of later faulting. Carboniferous, north of Bude, Devonshire, England (height of exposure about 6 m)

153. Boudinage of a quartz vein in mica schist. The sigmoidal shape of the boudinaged vein appears to be result of later deformation. Lower Paleozoic, near Cooma, New South Wales, Australia

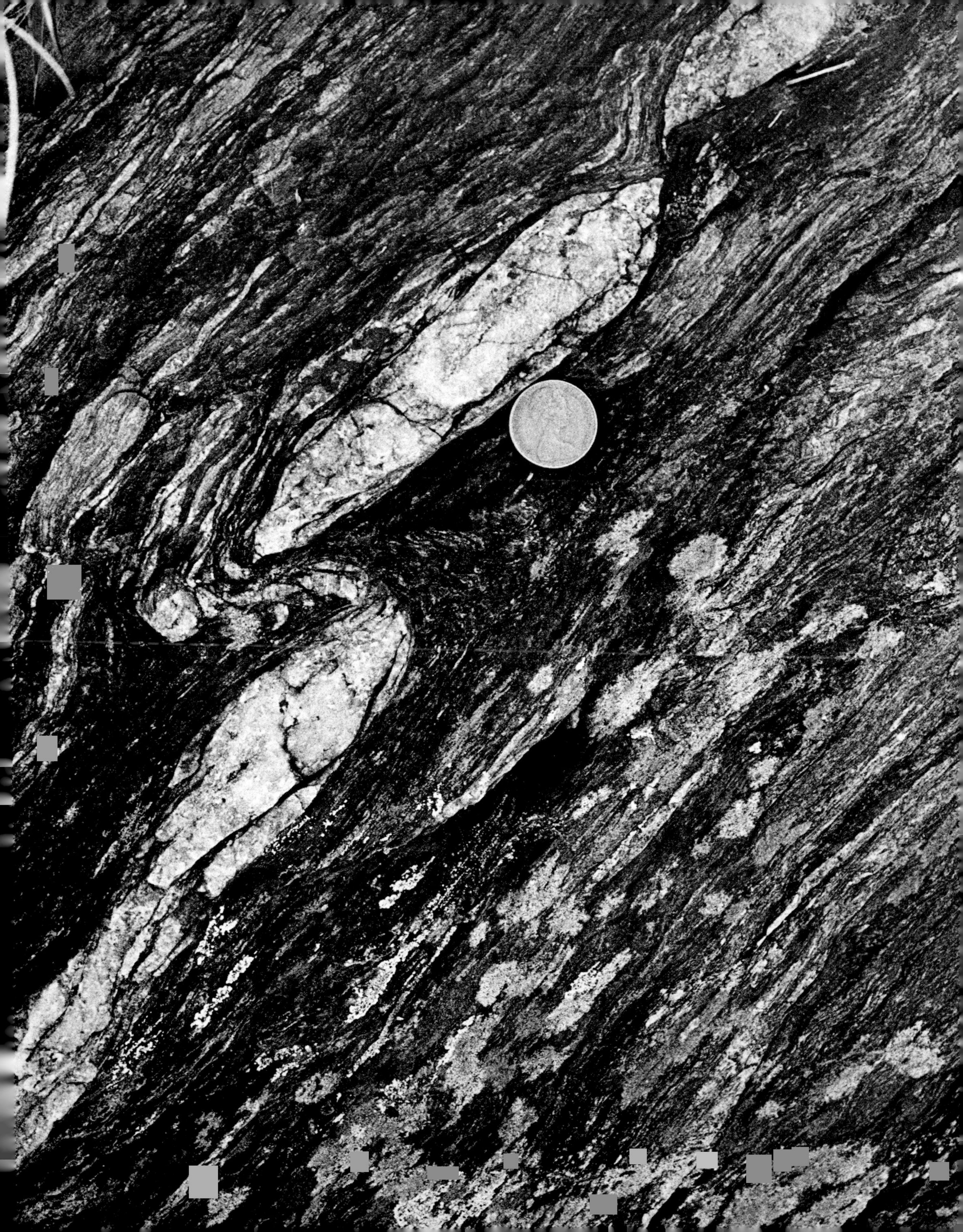

154. In schists with a prominent foliation thin oblique granitic veins (dipping left) contain both boudins and folds. The veins appear to have been longitudinally extended to form the boudins and then shortened again during subsequent folding which affects also the foliation of the schist. Paleozoic, near La Viella, Pyrenées, Andorra

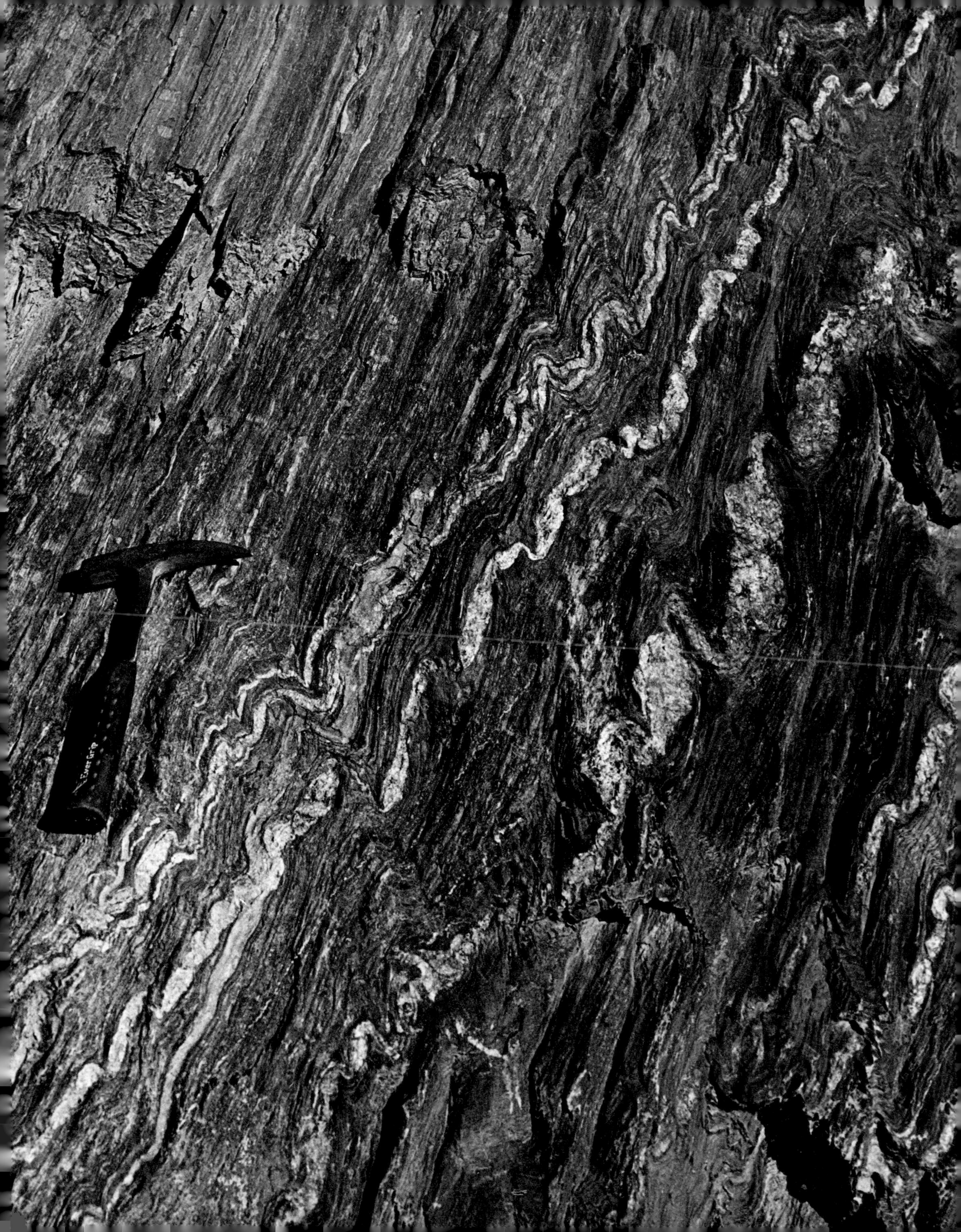
Ever Grip

155. On the underside of a boundinaged calc-silicate layer in marble a prominent lineation appears to mark the direction of extension during boudin formation. Lake Isabella, Sierra Nevada, California, U.S.A.

156A. Thin wedge-shaped quartz-filled fractures partially penetrate beds of massive graywacke. Carboniferous, near Welcombe, Devonshire, England

156B. Some of the veins are locally displaced by slip on bedding. From the same locality as Plate 156A

157. Parallel planar quartz-filled fractures in metamorphosed graywacke. Devonian, near Ilfracombe, Devonshire, England

158A. Massive graywacke-sandstone beds contain quartz-filled fractures at right angles to the beds. Fractures are not present in the intervening finely bedded shales. Carboniferous, Millook Haven, Devonshire, England

158B. Apparent "necking" and extension fracturing of massive sandstone bed. Individual fractures terminate at the margins of the bed. Carboniferous, Hartland Quay, Devonshire, England

159A. Quartz-filled fractures at right angles to bedding in massive graywackes are confined to one limb of a large fold. On this limb individual beds are thinned and the fractures appear to be extension fractures. Carboniferous, Millook Haven, Devonshire, England

159B. At the hinge of a small fold thinned beds on one limb are crossed by quartz-filled fractures whereas unthinned beds on the other (upper) limb are not. From the same locality as Plate 159A

160 A. Stout lenticular quartz-veins filled fractures in isolated beds of graywacke in shale. Jurassic, Le Lauzet, French Alps

160 B. Detail of lenticular quartz-veins. From the same locality as Plate 160 A

161 A. Quartz-vein displaced by slip on bedding surfaces. Moine, Glenshiel, Inverness-shire, Scotland

161 B. Curved quartz veins in bedded sandstones and shales. Carboniferous, Hartland Quay, Devonshire, England

Ever Grip
Ever Grip

162 A. Quartz-filled *en échelon* fractures in graywacke bed. Carboniferous, near Welcombe, Devonshire, England

162 B. *En échelon* fractures in graywacke bed. Col du Galibier, French Alps

163 A. Quartz-filled planar *en échelon* fractures in flaggy quartz-schist, north of Vågå, Oppland, Norway (height of exposure about 7 m)

163 B. *En échelon* fractures with sigmoidal shape are exposed on a bedding surface in graywacke. Near Sion, Valais, Switzerland (height of exposure about 30 m)

ALF R.

164A. Sigmoidal quartz-filled *en échelon* fractures in massive quartzite. Near Ringebu, Oppland, Norway

164B. Quartz-filled fractures in massive graywacke. Near Granville, Normandy, France

Ever Grip

Ever Grip

165 A. Quartz-filled *en échelon* fractures in massive schist. Moine, Kyle of Tongue, Sutherland, Scotland

165 B. Sets of intersecting quartz-filled fractures cross bedding in graywacke. Seyne, French Alps

166 A. Conjugate sets of quartz-filled *en échelon* fractures are exposed on the bedding surface of graywacke. Silurian, Spannlokket, Oslo Fjord, Norway

166 B. The *en échelon* fractures are arranged in two conjugate sets. From the same locality as Plate 166 A

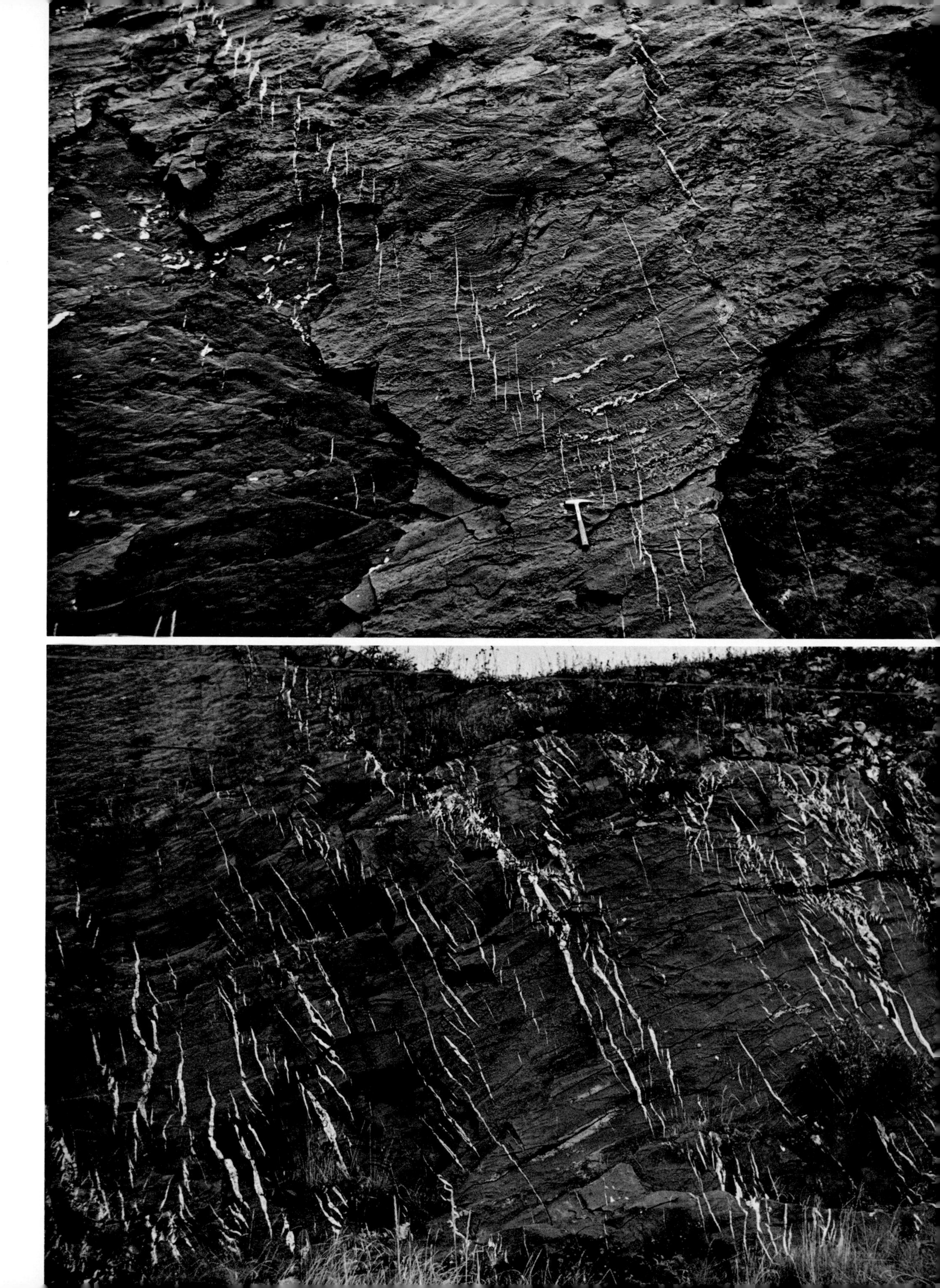

167. Detail of Plate 166 B

168. Extreme development of quartz-filled *en échelon* and other fractures in graywacke. Carboniferous, Petit St. Bernard, France

169. In calcareous graywackes two sets of veins obliquely inclined to bedding are present. One set of veins (vertical) is folded together with bedding. The other set (horizontal) remains approximately planar. Lower Paleozoic, near Trondheim, Sör-Tröndelag, Norway

Ever Grip

170 A. Isoclinally folded vein in gneiss. Near Tysfjord, Nordland, Norway

170 B. Quartzo-feldspathic vein obliquely inclined to foliation in gneiss is folded. Lower Paleozoic, near Korgen, Nordland, Norway

Estwing

Estwing

171. Folded vein in foliated gneiss is of approximately constant orthogonal thickness. From the same locality as Plate 170 B

172. Folded and boudinaged quartzo-feldspathic veins are obliquely inclined to foliation in schist. The folding, which affects also the foliation, appears to be later than the boudinage (see also Plate 154). Paleozoic, La Viella, Pyrenées, Andorra

MMOL
AVDA MERITXELL

173 A. Folded and unfolded sets of veins are present in deformed calcareous graywacke. From the same locality as Plate 169

173 B. Cross-cutting pegmatitic veins in gneiss. South of Fauske, Nordland, Norway

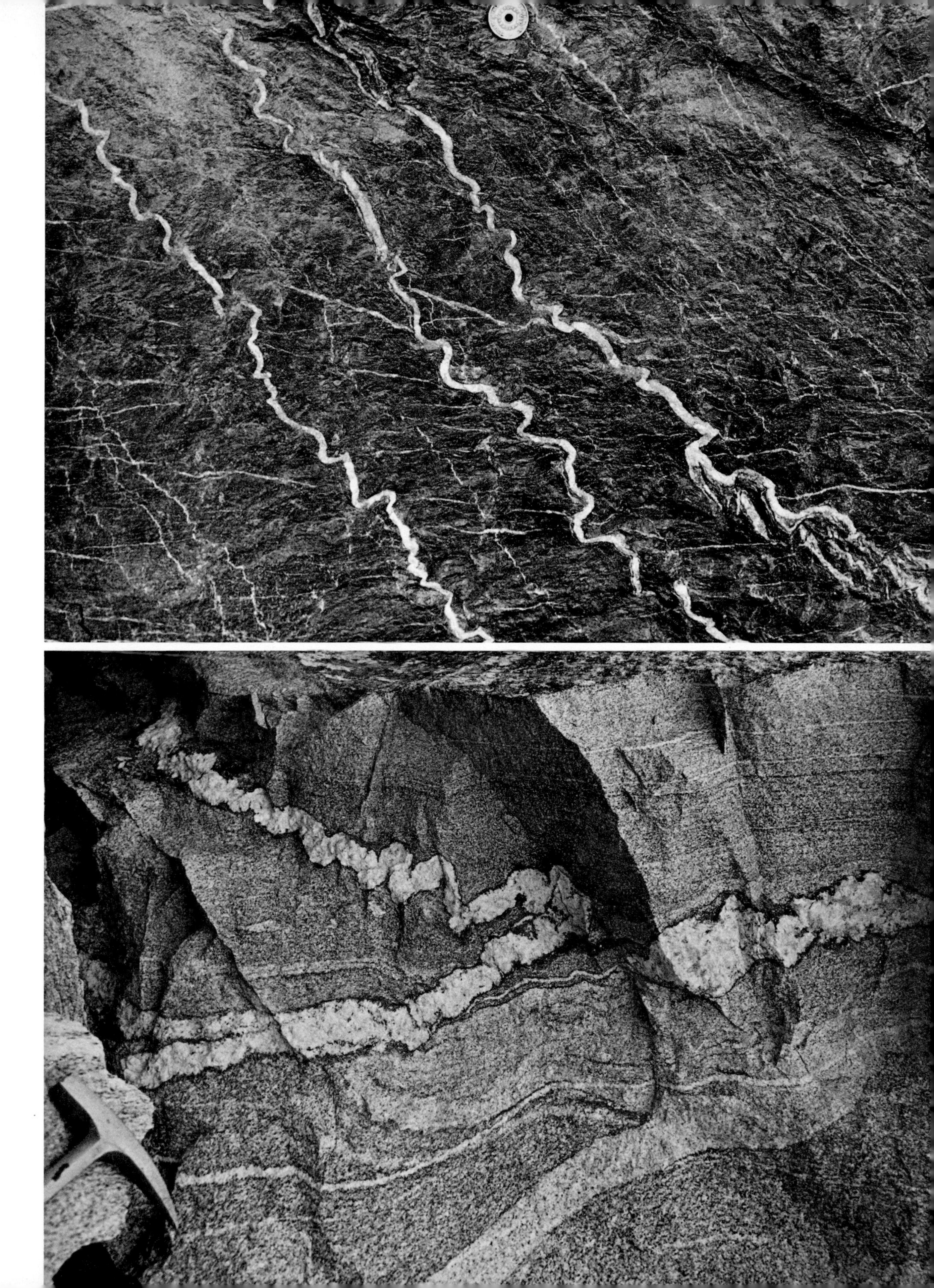

174 A. In well foliated gneiss veins parallel to foliation are unfolded whereas those obliquely inclined are tightly folded. Near Tysfjord, Nordland, Norway

174 B. Isoclinally folded quartz veins in schist. Moine, Strath Farrar, Inverness-shire, Scotland

175. Deformed quartzite-pebble conglomerate contains rod-shaped pebbles elongated in the direction of a regional lineation. Near Weldon, Sierra Nevada, California, U.S.A.

Estwing

176A. Deformed pebble conglomerate contains pebbles of very variable shape. All are elongated in the direction of a regional lineation and flattened in the plane of foliation of the enclosing mica schist. Pre-Cambrian, Panamint Range, California, U.S.A.

176B. Strongly elongated pebbles in deformed quartzite conglomerate. Bygdin, Oppland, Norway

NH14-45

177 A. Pebbles in a quartzite conglomerate are flattened in the plane of a regional foliation (dipping right). The plane of exposure is at right angles to a regional lineation which is also the direction of greatest elongation of the pebbles. Bygdin, Oppland, Norway

177 B. Strong elongation of pebbles is seen on an exposure surface parallel to regional lineation. From the same locality as Plate 177 A

178 A. In a deformed polymictic conglomerate pebbles of different composition are deformed by different amounts. East of Bergen, Hordaland, Norway

178 B. In a deformed conglomerate the light-colored fragments are boudins formed by extension of a granitic dike obliquely inclined to regional foliation. From the same locality as Plate 178 A

179A. Incipient folding of flattened pebbles in a deformed conglomerate by later movements. Bygdin, Oppland, Norway

179B. Flattened pebbles in a deformed conglomerate are bent by later folding. From the same locality as Plates 178A and 178B

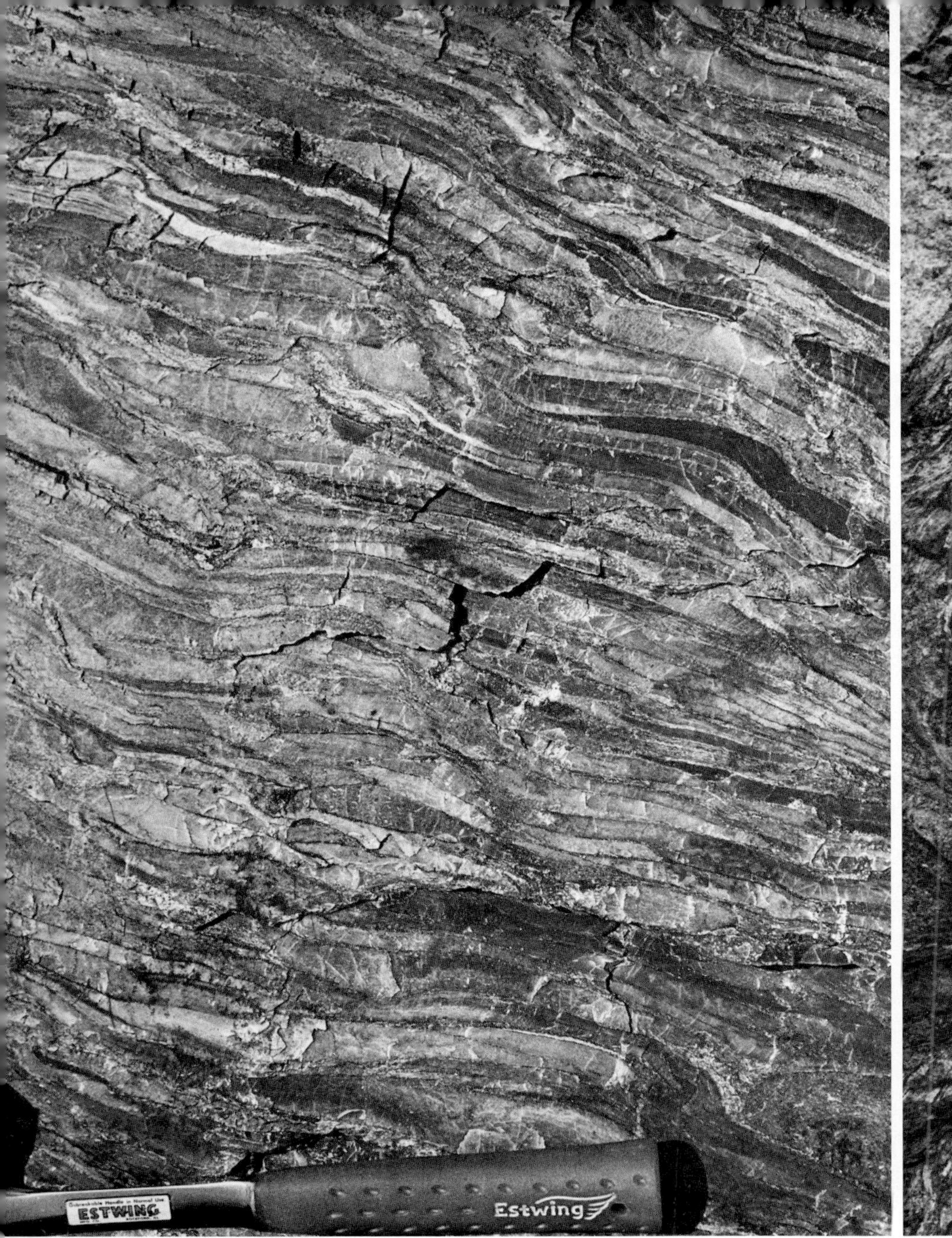
ESTWING
Estwing

180. In deformed sandstones cross-bedding is preserved in a distorted form. Bedding dips steeply left and the curved cross-bedding laminae are approximately horizontal. Moine, Glenshiell, Ross and Cromarty, Scotland

181 A. In deformed cross-bedded sandstones the cross-bedding laminae are almost at right angles to the main bedding. From the same locality as Plate 180

181 B. Cross-bedding laminae are locally inclined to the main bedding at very small angles (left). In other places (right) the main and cross-bedding laminae are folded together. From the same locality as Plates 180 and 181 A

182. Deformed bedding in metamorphosed sandstones. Moine, Strath Farrar, Inverness-shire, Scotland

EverGrip

183 A. Deformed trilobite *Olenus Cataractes* in shale. The direction of greatest elongation (horizontal) is marked by a pervasive lineation in the shale. Cambrian, Maentwrog, Merionethshire, Wales (width of fossil 3 cm)

183 B. Deformed trilobite *Asaphus Homfragi* in shale. The direction of greatest elongation (plunging gently right) is marked by a pervasive lineation and is obliquely inclined to the longitudinal axis of the fossil. The original bilateral symmetry of the fossil is destroyed. Cambrian, Penmorfa, Caernarvonshire, Wales (length of fossil 6 cm)

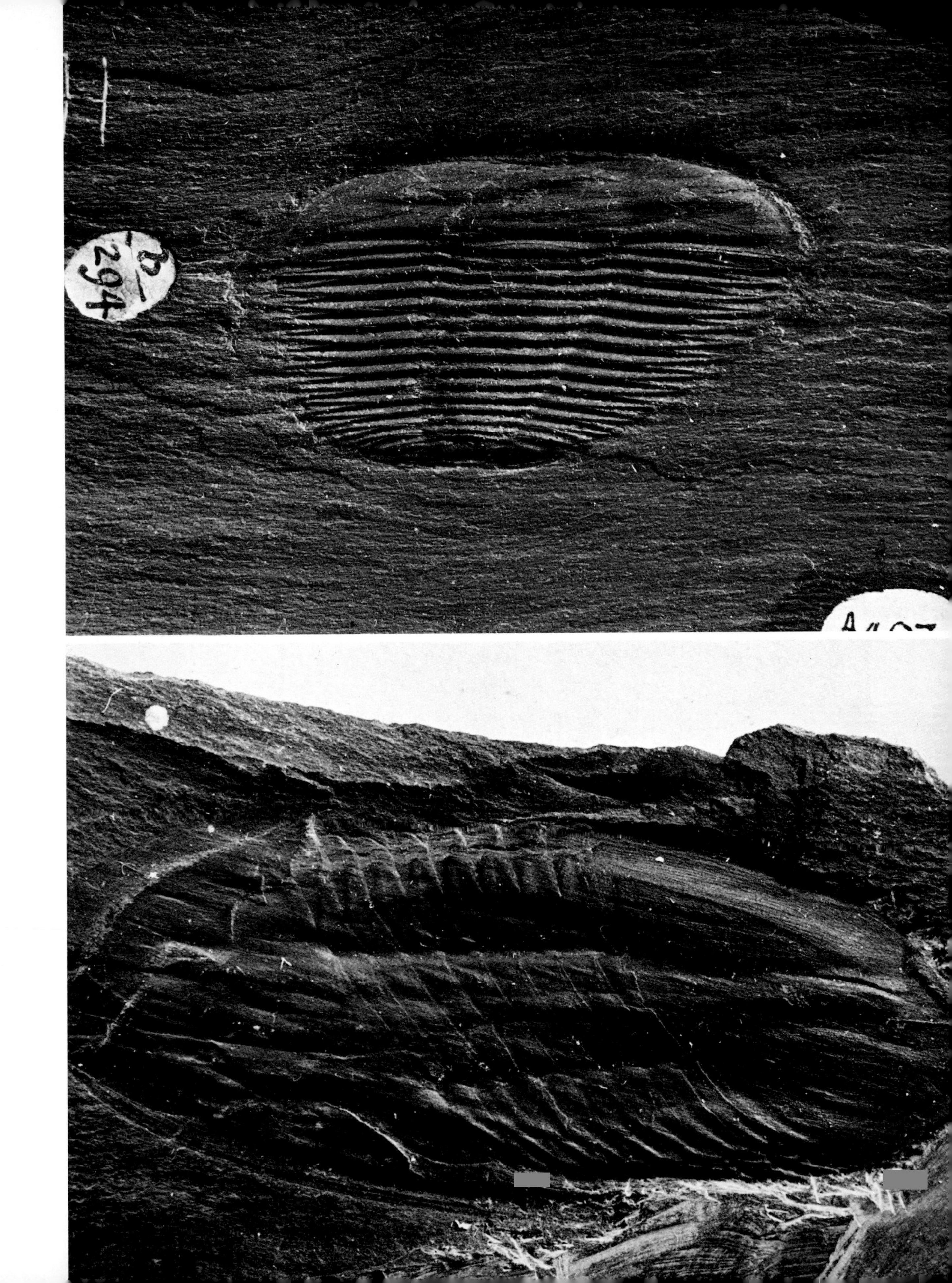

184A. Deformed trilobites in shale. The individual fossils are differently oriented in the bedding surface and are deformed by different amounts. Cambrian, Tremadoc, Caernarvonshire, Wales (width of specimen about 13 cm)

184B. Deformed trilobite *Angelina Sedgwickii* in shale. The lineation obliquely inclined to the long axis of the fossil is in the direction of greatest extension. Cambrian, Portmadoc, Caernarvonshire, Wales (scale is given by black and white squares of 1 cm side)

185 A. Deformed brachiopods *Lingulella Davisi* in shale. The shell impressions have been extended horizontally. Cambrian, Caernarvonshire, Wales (width of specimen 12 cm)

185 B. Weakly deformed lamellibranchs. Lower Silurian, Murrisk, Ireland (width of specimen 8 cm)

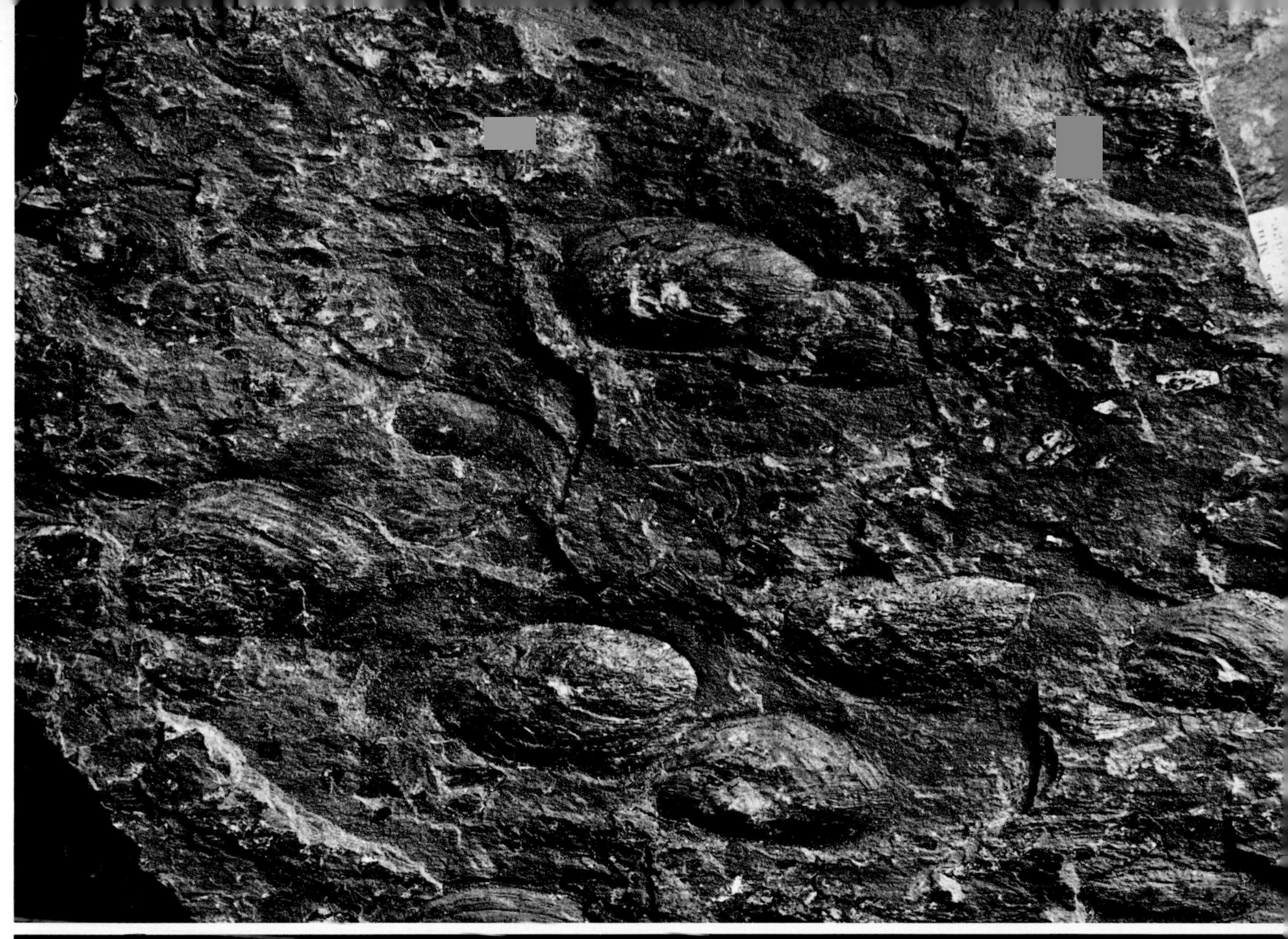

186 A. Graptolites with a variety of orientations in bedding surface of shale are deformed. Vertical dimensions of fossils are extended and horizontal dimensions reduced. Thus the horizontally oriented fossils tend to be short and broad whereas the vertical ones tend to be long and thin. Silurian, locality unknown (width of specimen 13 cm)

186 B. Deformed (right) and undeformed (left) goniatites *Euomphalus Pentangulatus*. Carboniferous, Dublin, Ireland

187A. An early lineation (horizontal) is crossed by a later faint crenulation (plunging left) on the foliation surface of a flaggy quartz-mica schist. Probably from the Oppdal region, Vågå churchyard, Oppland, Norway

187B. Two crenulation lineations on a foliation surface in schist. From the same locality as Plate 187A

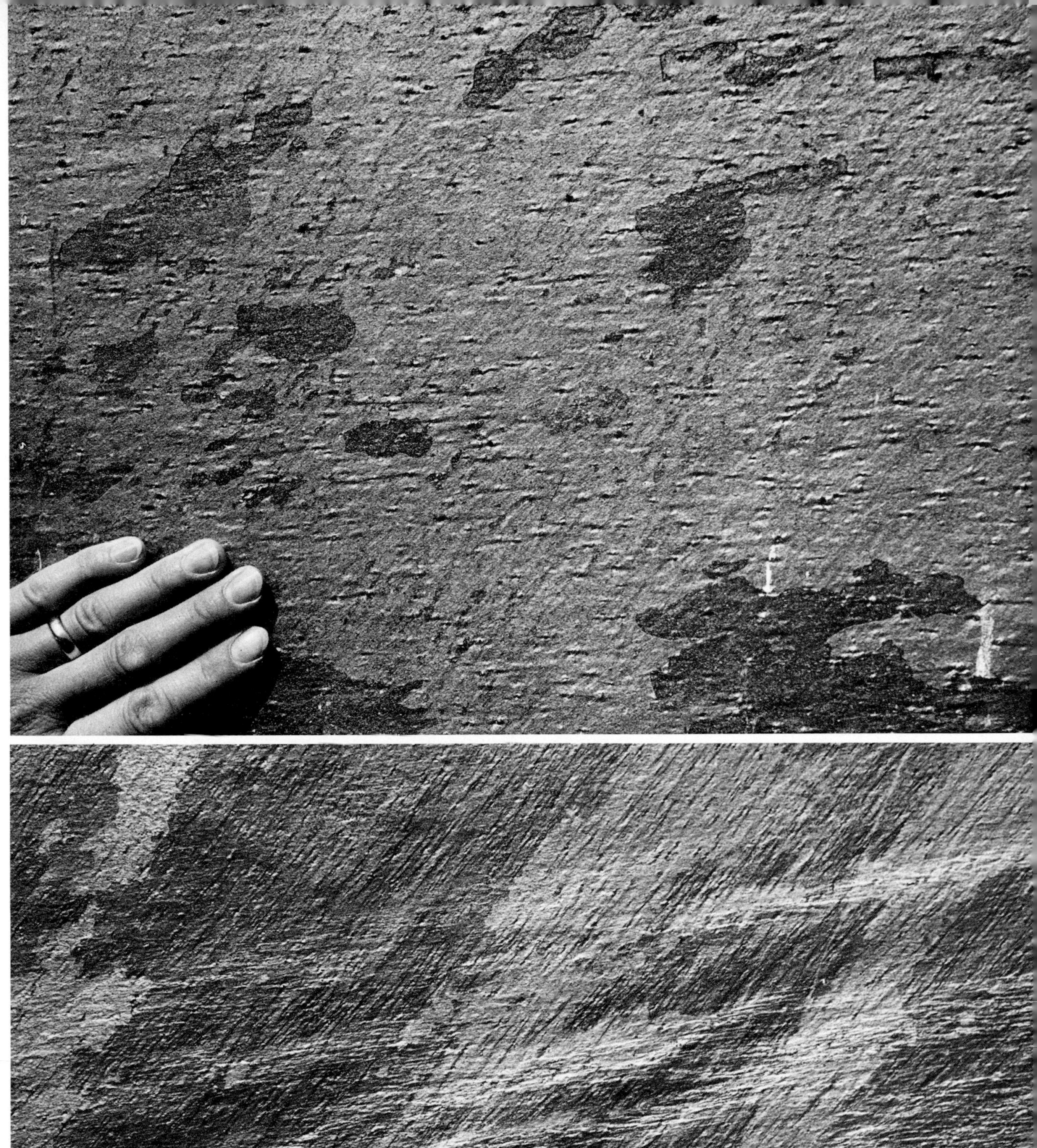

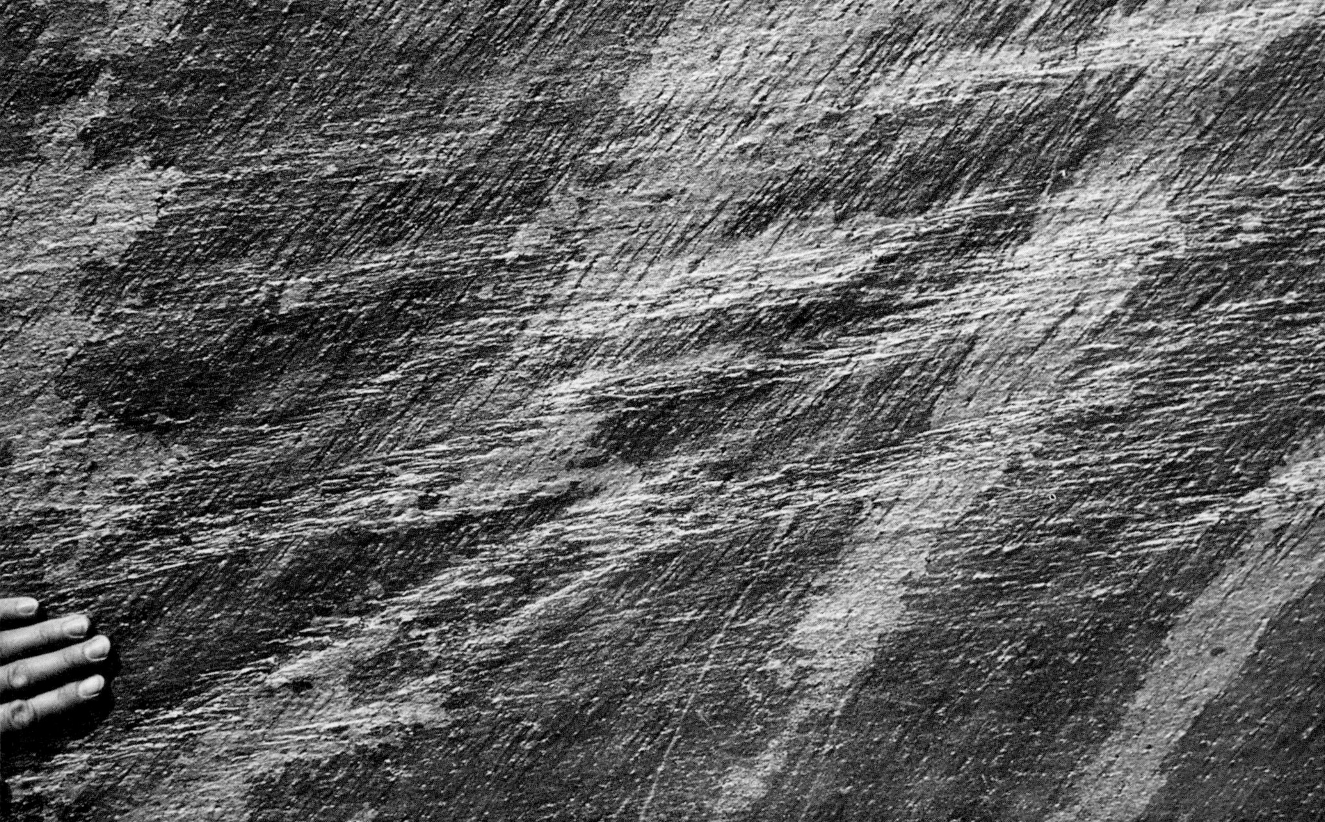

188 A. An earlier lineation (plunging steeply right) is defined by preferred orientation of elongated minerals and aggregates. A later small-scale folding with crenulation lineation (plunging gently left) distorts the earlier lineation. Lower Paleozoic, near Storlien, Sweden

188 B. Intersecting lineations in mica schist. Dalradian, Loch Leven, Inverness-shire, Scotland

189 A. An earlier lineation (plunging steeply right) in quartz-mica schist is crossed by a later intense crenulation lineation (plunging gently right). South of Lukmanier Pass, Ticino, Switzerland

189 B. A prominent lineation (plunging left) is distorted by later movements. The foliation surface in which the lineation lies is barely folded by the later deformation, but the lineation becomes markedly curved. Moine, Strath Farrar, Inverness-shire, Scotland

190 A. Early lineation (subvertical) is perpendicular to later folds and crenulations (plunging gently left) in mica schist. South of Lukmanier, Ticino, Switzerland

190 B. Subhorizontal crenulations and small folds in phyllite distort an earlier weak lineation (plunging steeply left). Lower Carboniferous, Morlaix, Brittany, France

191 A. An earlier pervasive lineation (plunging right) is distorted by later small folds (with axes parallel to hammer handle) in schist. Dalradian, near Tomintoul, Banffshire, Scotland

191 B. Wavy distortion of a prominent lineation in quartz-schist. Pre-Cambrian, near Karvik, Troms, Norway (height of exposure about 2 m)

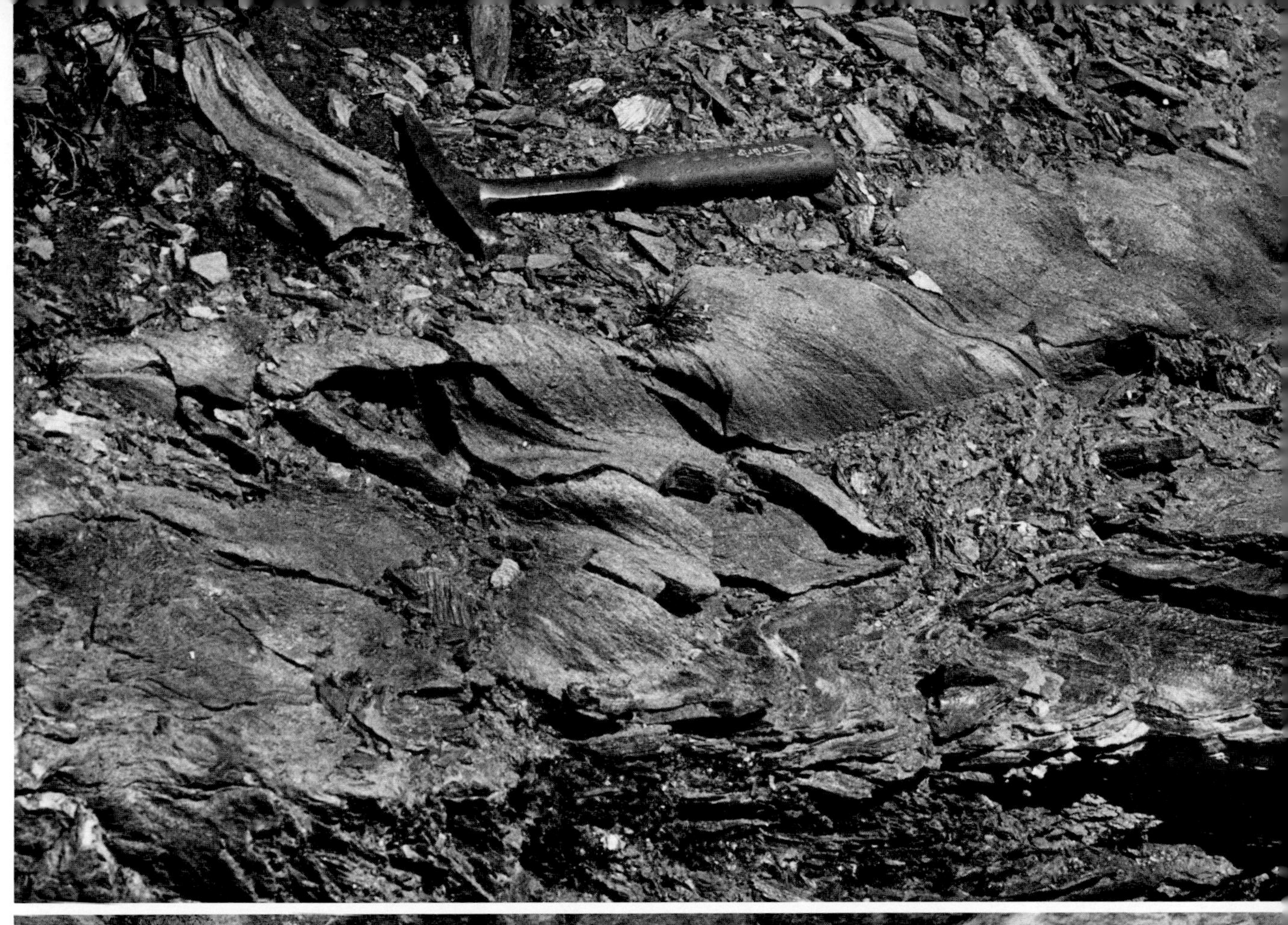

192. Earlier prominent lineation (plunging left) distorted by later cylindrical fold hinge (plunging steeply right). Moine, Strath Farrar, Inverness-shire, Scotland

193 A. In mylonitic gneiss a coarse lineation defined by rod-shaped mineral aggregates is bent by later folding. Near Musgrave Park, Musgrave Range, South Australia

193 B. Lineated quartzite lenticle in phyllite folded by later movements. Near Gjendesheim, Oppland, Norway

194. Layers of quartzite in mica schist contain earlier folds with axial surfaces dipping gently left. A later foliation (parallel to pencil) is developed in the schist and is parallel to the axial planes of sparse open folds developed on the limbs of the earlier tight folds (see, for example, the layer to the right of the upper end of the pencil). Dalradian, Loch Leven, Inverness-shire, Scotland

195. The trace of a later foliation is parallel to the hammer handle. Open folds with this surface as axial plane are present on the limbs of the earlier recumbent folds in the quartzite layers. From the same locality as Plate 194

196 A. In a schist relict hinges of tight folds with an earlier axial plane foliation are present. These are distorted by later folding and incipient development of foliation (dipping right). Lower Paleozoic near Dombås, Oppland, Norway

196 B. Earlier isoclinal folds in thin quartz layers in schist are distorted by later tight folds with subvertical axial planes. Paleozoic, near Åre, Sweden

Ever Grip

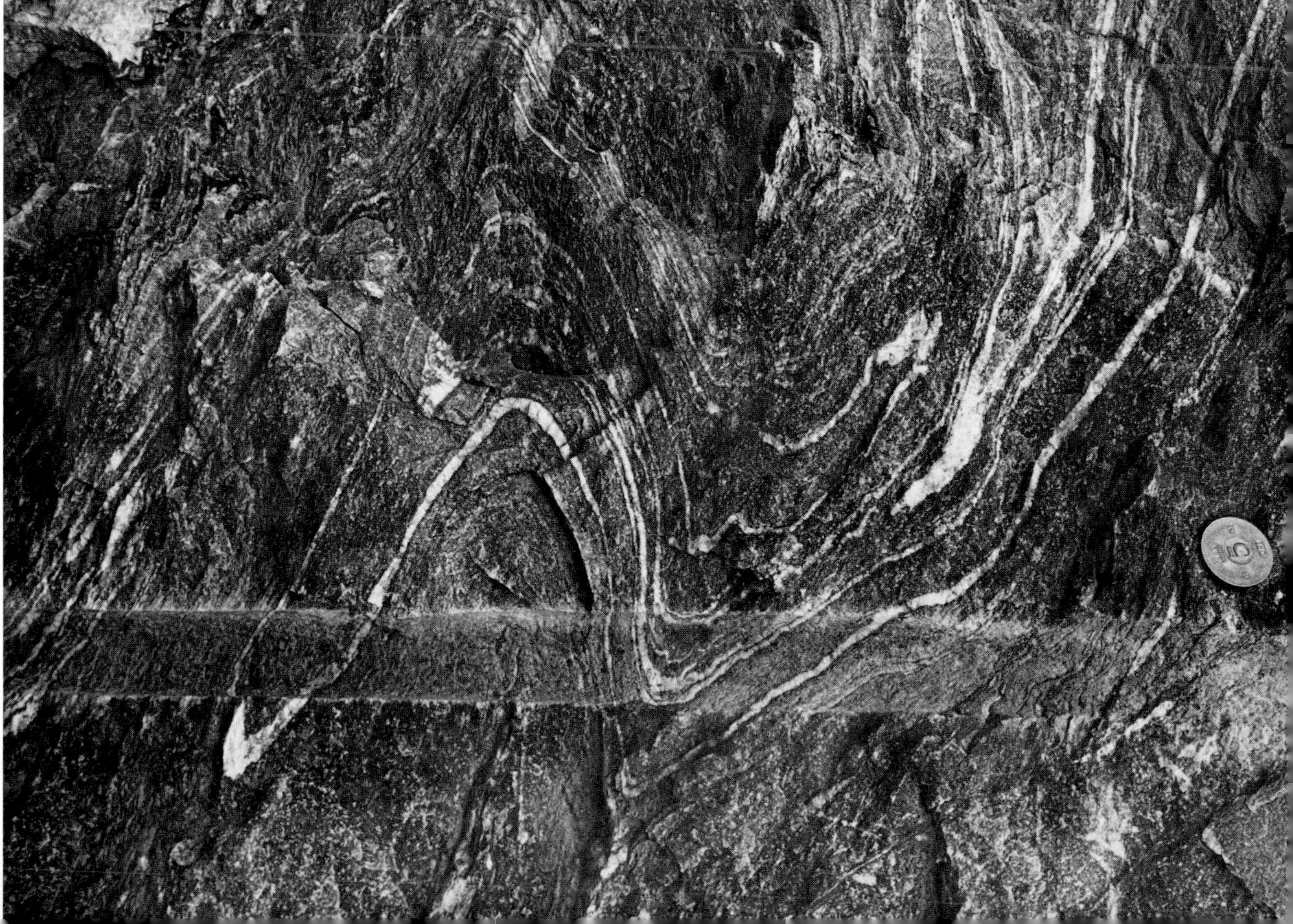

197A. In banded gneisses two fold generations are present. An earlier recumbent fold (above hammer) is distorted by later more open folds with horizontal axial surfaces. Pre-Cambrian near Evje, Aust Agder, Norway

197B. Detail of Plate 197A

198 A. Hinges of earlier small folds bent by later folding. Moine, Strath Farrar, Inverness-shire, Scotland

198 B. In flaggy quartz-schists earlier tight folds (upper center) are distorted by later more open folds with axial planes dipping right. Oppdal, Oppland, Norway (height of exposure about 6 m)

199. Flaggy quartz-schists containing early isoclinal folds (see, for example, upper center and left) are affected by later folding and mullion development about a subhorizontal axis. The later folds are very open and have horizontal axial planes. Oppdal, Oppland, Norway

200. Later folding of an earlier isoclinal fold in flaggy quartz-schist. From the same locality as Plate 198 B

201 A. In a thin layer of quartzite an earlier small fold with a horizontal hinge (center) is distorted and crossed by a later vertical foliation in enclosing mica schist. The lineation perpendicular to the fold hinge is thus later than the fold. Dalradian, Loch Leven, Inverness-shire, Scotland

201 B. Exposed bedding surfaces on graywacke sandstones contain an earlier horizontal lineation and weak folding, crossed by later open steeply plunging folds. Weak domes and basins result from the interference of the two fold sets. Near Sion, Valais, Switzerland

202. Dome and basin structure in quartzite layer affected by two generations of folds with different orientations. Dalradian, Loch Leven, Inverness-shire, Scotland

203. Ring-shaped outcrop formed by erosion of layers containing interfering fold sets. Moine, Strath Farrar, Inverness-shire, Scotland

Bibliography

The following books and papers have been selected for mention because of the influence they have had, or may yet have, on the development of ideas and techniques. The choice has been somewhat arbitrary–many of the papers listed could be replaced by similar works by the same or different authors.

BIOT, M.A.: Theory of folding of stratified viscoelastic media and its implications in tectonics and orogenesis. Bull. Geol. Soc. Am. **72**, 1595–1620 (1961).

CLOOS, E.: Oölite deformation in the South Mountain fold. Bull. Geol. Soc. Am. **58**, 843–917 (1947).

DENNIS, J.G.: International tectonic dictionary. Am. Assoc. Petrol. Geologists, Mem. 7, 196 p. (1967).

DIETRICH, J.H.: Origin of cleavage in folded rocks. Am. J. Sci. **267**, 155–165 (1969).

FAILL, R.T.: Kink band structures in the Valley and Ridge province, central Pennsylvania. Bull. Geol. Soc. Am. **80**, 2539–2550 (1969).

FLEUTY, M.J.: The description of folds. Proc. Geologists' Assoc. (Engl.) **75**, 461–492 (1964).

FLINN, D.: On folding during three-dimensional progressive deformation. Quart. J. Geol. Soc. London **118**, 385–433 (1962).

GREENLY, E.: The geology of Anglesey, vol. I. Geol. Surv. England and Wales, Mem., 388 p. (1919).

HANSEN, E.: Strain facies. Minerals, rocks and inorganic materials, monograph 2, 207 p. Berlin-Heidelberg-New York: Springer 1971.

HOBBS, B.: The analysis of strain in folded layers. Tectonophysics **11**, 329–375 (1971).

HOSSACK, J.R.: Pebble deformation and thrusting in the Bygdin area (southern Norway). Tectonophysics **5**, 315–339 (1968).

KNILL, J.L.: A classification of cleavages with special reference to the Craignish district of the Scottish Highlands. Report 21st Geol. Congress Copenhagen, part 18, 317–324 (1960).

MAXWELL, J.C.: Origin of slaty and fracture cleavage in the Delaware water gap area, New Jersey and Pennsylvania. Bull. Geol. Soc. Am. Buddington volume, 281–311 (1962).

PATERSON, M.S., WEISS, L.E.: Folding and boudinage of quartz-rich layers in experimentally deformed phyllite. Bull. Geol. Soc. Am. **79**, 795–812 (1968).

RAMBERG, H.: Natural and experimental boudinage and pinch and swell structures. J. Geol. **63**, 512–526 (1955).
RAMBERG, H.: Evolution of ptygmatic folding. Norsk Geol. Tidsskr. **39**, 99–151 (1959).
RAMSAY, J.G.: Superimposed folding at Loch Monar, Inverness-shire and Ross-shire. Quart. J. Geol. Soc. London **113**, 271–308 (1958).
RAMSAY, J.G.: Interference patterns produced by the superposition of folds of similar type. J. Geol. **70**, 466–481 (1962).
RAMSAY, J.G.: Structure and metamorphism of the Moine and Lewisian rocks of the north-west Caledonides. The British Coledonides, chap. 7, p. 143–175. Edinburgh: Oliver and Boyd 1963.
RAMSAY, J.G.: Folding and fracturing of rocks, 568 p. New York: McGraw-Hill Co. 1967.
ROERING, C.: Geometrical significance of natural *en échelon* crack arrays. Tectonophysics **5**, 107–123 (1968).
SITTER, L.U. DE: Boudins and parasitic folds in relation to cleavage and folding. Geol. Mijnbouw (NW. SER), **20**, 277–286 (1958).
TALBOT, C.J.: The minimum strain ellipsoid using deformed quartz veins. Tectonophysics **9**, 47–76 (1970).
TOBISCH, O.T.: The influence of early structures on the orientation of late-phase folds in an area of repeated deformation. J. Geol. **75**, 554–564 (1967).
TOBISCH, O.T., FLEUTY, M.J., MERH, S.S., MUKHOPADHYAY, D., RAMSAY, J.G.: Deformational and metamorphic history of Moinian and Lewisian rocks between Strathconon and Glen Afric. Scot. J. Geol. **6**, 243–265 (1970).
TURNER, F.J., WEISS, L.E.: Structural analysis of metamorphic tectonites, 545 p. New York: McGraw-Hill Co. 1963.
WATTERSON, J.: Homogeneous deformation of the gneisses of Vesterland, south-west Greenland. Medd. am Groenland **175**, 72 p. (1968).
WEISS, L.E.: Geometry of superposed folding. Bull. Geol. Soc. Am. **70**, 91–106 (1959).
WEISS, L.E., MCINTYRE, D.B.: Structural geometry of Dalradian rocks at Loch Leven, Scottish Highlands. J. Geol. **65**, 575–602 (1957).
WHITTEN, E.H.T.: Structural geology of folded rocks, 663 p. Chicago: Rand McNally and Co. 1966.
WILSON, G.: The tectonic significance of small scale structures and their importance to the geologist in the field. Ann. Soc. Géol. Belg. **84**, 423–548 (1961).

Subject Index

Numbers printed in **bold face** refer to the plates.